Robi Kremer
9D 1976-77

Tapestry of Eurasia

John Molyneux
Head, Geography Department
Harbord Collegiate
Toronto, Ontario

Eric Jones
Geography Department
Lawrence Park Collegiate
Toronto, Ontario

McGraw-Hill Ryerson Limited
Toronto Montreal New York London
Sydney Auckland Johannesburg Mexico Panama
Düsseldorf Singapore Kuala Lumpur New Delhi São Paulo

ISBN 0-07-082308-1

1 2 3 4 5 6 7 8 9 10 BP 5 4 3 2 1 0 9 8 7 6

Printed and bound in Canada
Cover design by Peggy Heath

Contents

EXAM
TERM ONE.

Preface

In this book we are trying to develop an understanding of certain key concepts in modern geography. The traditional pattern of teaching regional geography has usually hinged on the presentation of fairly comprehensive data in a well-organized framework composed broadly of physical features and human activities. By using the concept approach instead, we are trying to move away from the traditional pattern, and to develop in the students an awareness of some of the major aspects of modern geography.

The concepts we are attempting to develop are as follows:

Areal Differentiation, by which we hope to help the students understand why selected areas of the world are different not only from one another, but often also within themselves. To this end we use the whole of Eurasia, examined from the standpoint of climate, in chapter 1; the continent of Europe from the standpoint of farming, in chapter 5; the area of Scandinavia, looked at traditionally in chapter 6; the urban area of London, viewed from the standpoint of urban geography in chapter 8; and the nation of the U.S.S.R., from a traditional viewpoint, in chapter 16.

Change, in which we hope to show how selected countries and societies have changed in response to certain stimuli. We use the Arab world and the changes wrought by the development of petroleum as our example in chapter 12. The relatively new state of Israel is our example in chapter 13, while the much older nations of China and Japan are examined in chapters 17 and 18.

Global View, wherein we try to present a "bird's eye view" of a major topic. For this purpose we look at population throughout Eurasia in chapter 2 and then the process of urbanization in chapter 20.

Choice, whereby we hope to show the students that in the development of society, humanity often has a choice, and that the selection of a particular choice necessitates a certain course of action. Our main example is the European Community in chapter 4, but China and Japan in chapters 17 and 18 could also be used, because the changes in their societies are also largely the product of deliberate choice.

Regionalism, in which we aim to guide the students towards an understanding that certain areas of the world are marked by qualities that give them a special character. We use as our examples the Mediterranean in chapter 9 and the Indian sub-continent in chapter 15.

Resources, wherein we attempt to show what constitute resources and how they fit into human activity. For this purpose we have chosen energy resources and related them to economic growth in chapter 11.

Spatial Interaction, by which we intend the students to gain an understanding of how the different parts of the world are linked together. We therefore study transportation in chapter 3 and migrations in chapter 14, both of them throughout Eurasia.

Support of Life, whereby we try to analyze the different ways in which people manage to gain a living. In chapter 7 we look at how the Swiss manage to earn a high standard of living, while in chapter 10 we examine countries whose people have great trouble gaining a living. In chapter 19 we investigate how a particular resource — fish — can be used to help support human life.

While we know that our selection of topics is not necessarily a universal choice, we are confident that it offers enough to be interesting and more than enough to be useful.

JOHN MOLYNEUX
ERIC JONES
Toronto 1976

1. CLIMATE

AREAL DIFFERENTIATION

An Italian farmer grumbles as he gazes over his brown fields in August. He would like a good thunderstorm to make his land fresher and his crops greener. Even just a little shower would greatly improve the year's vine crop, he feels.

Tennis is being played at Wimbledon, England. From the 4 corners of the earth, professionals and amateurs have gathered to play the ancient game. An umpire on the centre court looks skyward in apprehension. A native of Ottawa, he remembers his mother laughing about British summer weather, how unsettled it can be. On go the covers as muttering players, officials, and spectators endure a feature of Britain's maritime climate. Often gentle but always infuriating, rain is falling once again.

On the oppressively humid landscape of the Ganges delta, a Bengali peasant starts in alarm. His irrigation channels, mere scratches in the earth, have water in them. That cannot be, he thinks, for who but he would decide when to water the rice? And why is the flow of water so great? As he watches, the level of the flowing stream inches upward and then overflows the bank.

He joins a mad stream of humanity as it flows like a torrent from a burst dam. As the people rush into their village, the setting sun behind them glints redly from flooded fields. While the panic-stricken chatter from the people reaches fever pitch, great clouds roll up from the south, preceded by a stiffening wind that begins to tear at the palm trees. Then comes a downpour of heavy rain —blotting from sight objects less than 100 m away. The sun sets, unnoticed. The wind builds to a shrieking fury, hurling rain that seems to emanate from a million high-pressure hoses. Homes collapse and their occupants are blown away. At speeds in excess of 250 km/h, hurricane winds are irresistible. They can even blow away people who are lying prone, head and shoulders to the wind. On the dreadful night of November 12-13, 1970, half a million people were drowned in Bengal when a great wave of seawater was blown into the Ganges delta by a tropical cyclone. The river backed up and the inhabitants of the region were caught by rapidly rising water. They had no high land on which to take refuge and drowned in a scene of frightening destruction.

The effects of climate upon humanity are innumerable. In western Europe, weather and its vagaries are usually treated as annoyances. Rain spoils a game of soccer or golf, catches people unawares at the bus-stop, or is used as the most common way of starting a conversation. Only rarely does the weather bare its teeth in fury. However, when it did so in 1953, a great storm in the North Sea drowned hundreds of people in half-a-dozen countries and inundated hundreds of square kilometres of good agricultural land with salt water. A great deal of this land never completely recovered.

By way of contrast, people in Asia treat the weather with a great deal more respect. The great proportion of Asia's teeming millions exist using a fairly low level of technology. Thus, most Asians are agricultural workers or merely simple peasants. Their lives are keyed to the seasons in such a series of strong bonds that many people still worship a number of gods in the hope of protection from natural disaster.

While the influence of weather upon the works of people is usually direct and fairly obvious, weather also affects people in ways that are only now being isolated and studied.

Some of these ways are easily observed and recorded. Who has not felt lethargic when it is humid and hot? The person who "feels the weather in my bones" is speaking the literal truth. Rheumatism and arthritis *do* worsen in the low pressure that precedes a storm, and old wounds are aggravated. There can be no doubt that the sight of green leaves and growing plants gives a psychological lift to the spirit. "In spring, a young man's fancy turns to thoughts of . . . ." Precisely the same feelings of elation are felt in Burma when the wet monsoon bursts over the land and gives renewed hope of life for at least another year. The science of *meteorpsychiatry* seeks to explain the ways in which these and other natural phenomena influence human beings' psychological behaviour.

The word *weather* refers to short-term changes in atmospheric conditions — rainstorms giving way to sunshine, for example. *Climate* is the long-term variation in weather. By short-term, we mean a period as long as a month or as short as an hour. By long-term, we imagine a period of time spanning as much as a year or more, or perhaps only a season. Do not imagine that there are only

An "ill wind that blows nobody any good" is called the *fohn* in Germany and Switzerland, and the *sharav* in Israel. Both are dry winds that stimulate the brain to over-produce serotonin, a chemical essential for its proper functioning. This results in depression, aggravating problems of serious crimes, epileptic fits, and mental disorders.

Many parts of the world have weather lore in verse form.

Ants that move their eggs and climb, rain is coming anytime. — India, Japan.
Shrill calling cranes flying high and slow, a pleasant fall we shall know.— U.S.S.R.
Spring without rain: abundant grain. A dry fall: no grain at all. — China.
The evening red and the morning grey are the tokens of a bonnie day.— Europe.

The world's heaviest hailstone weighed 1.9 kg. It fell in Kazakhstan, U.S.S.R.

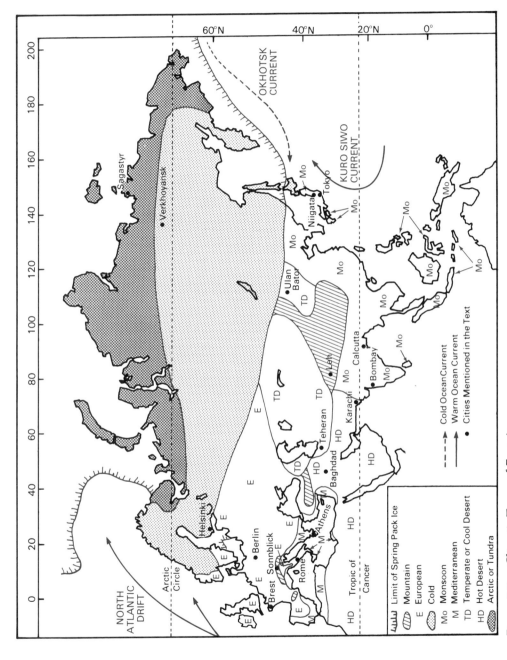

Figure 1-1. Climatic Types of Eurasia

4 seasons! Seasons can be hot, warm, cool, or cold; dry or damp; windy or still and long or short — *besides* being spring, summer, fall or winter!

A part of the world that has a predictable climate is said to show a *climatic type*. For many years geographers have puzzled over what to include when they give a definition of a climatic type. As you can imagine, there have been many attempts at describing the climates of the earth as completely as possible, and not one of them is the same in every detail. The problem is rather like the one that would face a person told to define every person in the world. How is this to be done? Which is most important out of a list of check-points such as height, weight, race, sex, colour of hair, colour of eyes, colour of skin, religious beliefs, country of origin, political beliefs, etc.? Of course, everybody fits in somewhere, but isn't it easier to talk in terms of groups so that although there can be quite a bit of variation *within* a group, the variation *between* groups is even greater? Climates are often treated like that. Although there is a lot of difference between the climates of India and Japan, they both still have climates that belong to the *monsoon* variety. Thus, both countries rely greatly on rice as a staple cereal.

If you compare the map shown in Figure 1-1 with any other climate map of Eurasia that you can find, you will probably discover that it differs in some details — but only in details. The main areas of each climate type will be roughly the same. Only the actual boundaries may be different. This is understandable, for the climates in Figure 1-1 do not have sharp boundaries at all. The boundaries are really rather fuzzy edges where one type of climate gives way to a different one. Different people have different ideas about just where monsoon climates give way to cold climates, or deserts become mountain climates, and so on.

The world's highest recorded air pressure was 1083.8 mb in Agata, U.S.S.R.

MEDITERRANEAN CLIMATES

Hot, dry summers and cool, moist winters characterize this type of climate. It is usually associated with the cultivation of the olive tree and the grape vine. These plants, along with the cork oak, are known as *xerophytes*. This means that they have the ability to withstand long periods of drought. They are physically adapted to the Mediterranean summer. Their bark is thick and insulating

	J	F	M	A	M	J	J	A	S	O	N	D	
Rome	7	8	11	14	18	22	24	24	22	17	12	8	temperature (°C)
	81	69	74	66	56	41	18	25	64	127	112	99	precipitation (mm)
Athens	8	8	11	15	19	23	27	27	23	19	14	11	
	51	43	30	23	20	18	8	13	15	41	66	66	

while their leaves are short, narrow, and shiny to reduce the loss of water through transpiration.

For most if not all Mediterranean countries with this type of climate, tourism in the summer season is an important part of the economy. Millions of people from the cooler lands to the north descend in summer to soak up the sun, lie on the beach, and just generally relax.

HOT DESERTS

Annual rainfall totals here are less than 250 mm. There is usually no distinct rainy season. When precipitation comes at all, it is often at long intervals and is quite violent. The reason for the extreme *aridity* is found in the direction of the prevailing wind. Coming from the great Asian land-mass to the northeast, such winds are usually extremely dry. Vegetation is sparse. Traditionally humans have responded to such an environment by becoming nomadic herders of sheep, goats, and camels.

	J	F	M	A	M	J	J	A	S	O	N	D	
Baghdad	9	12	16	22	27	32	35	35	31	27	17	2	temperature (°C)
	30	33	33	23	5	—	—	—	—	3	20	30	precipitation (mm)
Karachi	18	20	24	27	29	31	29	28	27	27	24	19	
	13	13	10	5	3	3	74	38	13	—	3	3	

COOL DESERTS

With a similar lack of precipitation to the hot deserts, cool deserts lie farther north and experience the full severity of interior continental winters. At that time of year, winds are both cold and outblowing; this is hardly calculated to give them much precipitation. Summers can be almost as hot as those of the hot deserts. Daily temperatures of over 38°C are not uncommon. Monthly averages are lower than this though.

	J	F	M	A	M	J	J	A	S	O	N	D	
Teheran	1	6	9	16	22	27	29	28	25	18	11	6	temperature (°C)
	40	25	49	36	13	3	5	—	3	8	25	33	precipitation (mm)
Ulan Bator	−27	−20	−11	1	9	14	17	14	9	−1	−13	−22	
	—	3	—	—	8	43	66	53	13	3	3	3	

EUROPEAN CLIMATES

Within this climatic type, there is a good deal of variation. On the coastal margins of the west, the *maritime* type of climate is found. Here rainfall is of the relief or orographic type and tends to have a maximum in winters. The winters themselves tend to be relatively mild compared with the interior *continental* type of climate. There, summers are hotter than the maritime type but have rainfall at its highest then, owing to the activity of thunderstorms, which bring convectional rain.

In November 1972, 63 people were killed and hundreds injured when hurricane-force winds swept across Europe.

	J	F	M	A	M	J	J	A	S	O	N	D	
Brest	7	7	8	10	12	15	18	18	16	12	10	8	temperature (°C)
	62	58	53	50	58	36	31	46	60	82	74	89	precipitation (mm)
Berlin	-1	0	3	9	14	17	19	19	14	9	3	0	
	41	34	38	36	46	55	72	55	41	41	41	46	

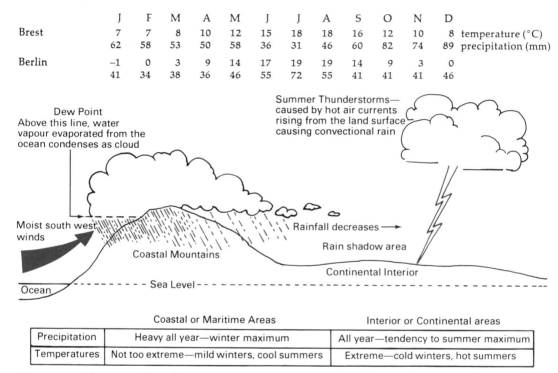

	Coastal or Maritime Areas	Interior or Continental areas
Precipitation	Heavy all year—winter maximum	All year—tendency to summer maximum
Temperatures	Not too extreme—mild winters, cool summers	Extreme—cold winters, hot summers

Figure 1-2. European Type Climates

COLD CLIMATES

Save for a narrow strip along the Norwegian coast, which is washed by the North Atlantic Drift as can be seen in Figure 1-1, winter temperatures here often fall below 0°C. Away from the coastline, temperatures are some of the coldest of the world in winter. In Siberia, for example, −50°C is not uncommon. The air, though, is still and dry and the sun shines often. It is only when blizzards blow that temperatures feel below freezing. For many decades (before research in Antarctica) the lowest temperature ever recorded in the world was taken in Siberia at Ver-khoyansk. It was −68°C. This compares with Canada's lowest ever recorded temperature, −62°C at Snag in the Yukon.

Summers are often very hot, but precipitation is low, usually less than 700 mm annually and often less than 500 mm annually.

	J	F	M	A	M	J	J	A	S	O	N	D	
Helsinki	−6	−6	−4	1	8	14	17	16	11	6	0	−4	temperature (°C)
	43	34	34	34	43	43	53	70	60	62	60	58	precipitation (mm)
Verkhoyansk	−50	−44	−32	−13	2	13	16	11	2	−14	−37	−46	
	5	2	2	5	7	22	24	24	12	10	7	2	

ARCTIC CLIMATES

The southern part bears the familiar name of *tundra*. Here the almost permanently frozen ground relents for the short, cool summer of 3 months and permits mosses, lichens, and coarse grasses to grow. In sheltered locations, such as along river banks, dwarf willow, birch, and alder are not uncommon.

Beyond the Arctic Circle, the ground is frozen permanently. Little grows. Everywhere here there is at least one day when the sun is visible at midnight. This culminates in the 6-month periods of light and darkness at the North Pole itself. Even with 24 hours of sun, however, the warmth of an Arctic day is not great because the sun is so low in the sky.

You cannot count on averages. In July 1972, temperatures rose to over 32°C north of the Arctic Circle, starting forest fires in Norway. Hundreds of kilometres to the south, in Frankfurt, Germany, 4 people died from heatstroke.

	J	F	M	A	M	J	J	A	S	O	N	D	
Spitzbergen	−16	−19	−19	−13	−5	2	6	4	0	−6	−12	−14	temperature (°C)
	34	31	26	22	12	10	14	22	24	29	24	36	precipitation (mm)
Sagastyr	−37	−38	−34	−22	−9	0	5	3	1	−14	−27	−33	
	2	2	—	—	5	10	7	34	10	2	2	5	

MOUNTAIN CLIMATES

As you might have imagined from the name, this type of climate is dominated by altitude. Because the air is very thin, the sun's rays are quite fierce, particularly in the infra-red and ultra-violet wavelengths. Shadows are bitterly cold. Because the air is so cold, it tends to sink and blow down the deep mountain valleys to lower altitudes.

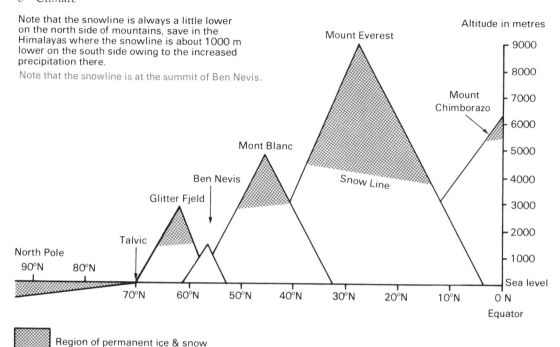

Note that the snowline is always a little lower on the north side of mountains, save in the Himalayas where the snowline is about 1000 m lower on the south side owing to the increased precipitation there.

Note that the snowline is at the summit of Ben Nevis.

Region of permanent ice & snow

Figure 1-3.   The Altitudes of the Snow Line

As it does so, this *katabatic* wind can increase in temperature by 15°C or more. It is likely to produce fire warnings because it is so dry, but in fall it is welcomed in many regions because it is a fine aid to ripening crops.

Since the zone of maximum precipitation anywhere on the surface of the earth is usually at or below about 1800 m, mountain climates are often quite dry. The snowline is found in mountain climates, but its altitude depends on latitude.

An interesting effect of altitude upon man is the appearance of mountain sickness at heights above sea level in excess of 4500 m. Above 6000 m, mountaineers require oxygen to supplement the thin air their heaving lungs strain for. Mount Everest (8850 m) was only conquered first in this way.

Japan Information Centre
Consulate General of Japan, Toronto

The snow of winter looks much the same anywhere in Eurasia.

	J	F	M	A	M	J	J	A	S	O	N	D	
Leh	−8	−7	0	6	10	14	17	16	12	6	0	−6	temperature (°C)
	10	7	7	5	5	5	12	12	7	5	—	5	precipitation (mm)
Sonnblick	−13	−13	−13	−9	−4	−1	1	1	−1	−5	−9	−12	
	118	118	151	158	149	132	134	122	110	122	110	127	

MONSOON CLIMATES

The monsoon is a seasonal wind. The *dry monsoon* blows outward from the Asian land-mass in winter. Precipitation is low and skies often a bright blue. After January, temperatures rise unpleasantly until June, when the *wet monsoon* starts to blow. The wet monsoon comes from the sea, bringing heavy rain all through the summer until October or November, in some cases. Some 80% of the annual rainfall is brought by the monsoon; in India, the highest precipitation totals in the world are produced on or near the Himalayan mountains. Cherrapunji (1300 m), for example, receives in excess of 11 500 mm annually. This works out to in excess of 30 mm daily; the daily average for southern Ontario is a mere 2 mm!

Probably the most significant climate type in Eurasia, the monsoon affects 4 500 000 km^2 (nearly 5% of the world's land area) but a staggering total of 2.5 billion people, or some 60% of the world's population. Without this heavy summer rain, many millions of these people would starve to death, for the monsoon nurtures rice, a staple cereal and the major part of many Asians' diets.

India has a population of over 600 000 000 people. The majority are subsistence farmers who rely greatly on the rain of the monsoon to provide enough moisture. Arriving first at the southwest coast of Malabar on June 3, the monsoon reaches all India by the end of the month.

The monsoon rains came one month early for Bangladesh in 1974. They hit in early July, causing floods that killed over 2000 people and did $3.5 billion damage. Half the land area was covered; cholera appeared; and the U.S.A. was asked for 500 000 t of grain. A world food shortage resulted in the U.S.A. offering only 200 000 t of grain.

Bogor, Indonesia, averaged 322 thunderstorm days in each year between 1916 and 1919.

	J	F	M	A	M	J	J	A	S	O	N	D	
Bombay	24	24	27	28	30	29	27	27	27	28	27	25	temperature (°C)
	3	3	—	—	18	505	610	368	269	48	10	—	precipitation (mm)
Calcutta	19	22	27	29	30	29	29	28	28	27	23	19	
	10	25	36	56	142	302	323	348	254	124	15	5	

Japan is remarkable for the fact that it has 2 wet monsoons. Besides the usual summer (wet) monsoon, the winter monsoon crosses the Sea of Japan and brings a second precipitation maximum to the west coast at that season.

	J	F	M	A	M	J	J	A	S	O	N	D	
Tokyo	3	4	7	13	17	21	24	26	22	16	11	5	temperature (°C)
	56	71	112	124	145	165	135	145	221	188	107	53	precipitation (mm)
Niigata	2	2	4	11	15	19	23	26	21	15	9	4	
	185	348	98	101	89	125	149	125	178	137	173	218	

The end of the summer monsoon in October and November is a season of danger throughout monsoon

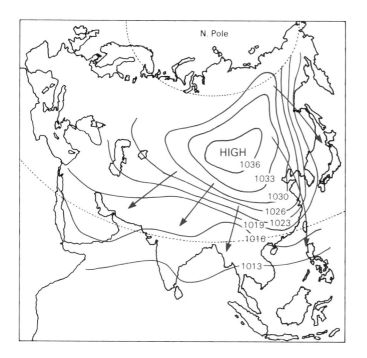

Pressure given in millibars. Note the cold, outblowing winds (⤡).
High pressure air is cold, therefore it sinks.

Figure 1-4 (a). Eurasia — January Pressure and Winds

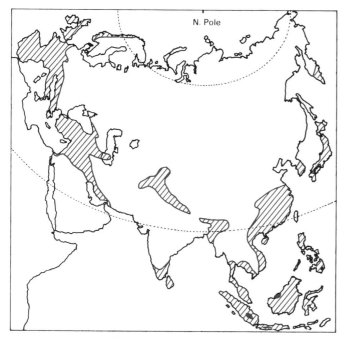

over 2000 mm

over 250 mm

Figure 1-4 (b). Eurasia — Rainfall November 1 to April 30

Shell Photographic Service

Before the monsoon breaks, the land seems dry and lifeless. When the rains come, the track in the foreground will become a bog, impassable to wheeled vehicles. The trees and shrubbery will quickly become leafy, and this part of Northeast Thailand will be cloaked in jungle again.

Japan Information Centre
Consulate General of Japan, Toronto

The monsoon rains mean that rice cultivation can start once more.

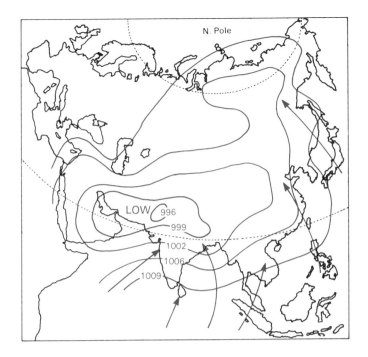

Pressure given in millibars. Pronounced low pressure over Pakistan draws in damp air from the ocean ().

Figure 1-4 (c).   Eurasia — July Pressure and Winds

Japan Information Centre
Consulate General of Japan, Toronto

At the end of the monsoon, rainfall drops sharply, the fields dry, and the rice harvest may commence. In some parts of monsoon Asia, two rice crops are obtained each year.

over 2000 mm

over 250 mm

Figure 1-4 (d).   Eurasia — Rainfall May 1 to October 31

Asia. This is the time of year when tropical *cyclones* can strike. Known as *typhoons* in the China Sea, they claim many, many lives each year and do untold damage. One has already been described in this chapter.

Apart from the bodily and mental effects of weather and climate on people, there are other, more indirect effects. To a certain extent, climate determines soil type. To a greater extent, climate also determines the growth of plants.

Figure 1-5 illustrates in a general way how the vegetation that grows naturally varies over the face of Eurasia. To a great extent it is the amount of usable or available moisture that determines what the natural vegetation of any area is to be.

High rainfall areas tend to have dense forest cover, such as the *Scandinavian forest.*

Low rainfall areas tend to have extensive grasslands, such as the *steppe.*

Where precipitation is very low, such as in a *hot* or *cold desert*, little or no vegetation survives, save for cacti, acacia bushes, and other drought-resistant shrubs. Can you remember the correct name for such plants? The word has been used earlier in this chapter.

It is obvious that, in a hot desert, what little rainfall there is will quickly evaporate into the air. Hot air at a temperature of 30°C can hold at least 6 times as much water in the form of vapour as can air at a temperature of 0°C. What is perhaps not so clear is that precipitated moisture in a polar desert, while it does not evaporate very easily, is still not available in very large quantities for plant growth. This is because most of it is frozen. As far as the mosses and lichens of the *tundra* are concerned, their yearly ration of moisture is just about the same as that of the *hot desert*, which is thousands of kilometres away!

As a general rule, temperature affects growing plants; if the temperature falls to 6°C, they stop growing. Usually they remain dormant — as in the great northern forests — until the temperature has risen sufficiently for them to resume growth. Thus if the mean or average temperature in a month is above 6°C, then that month can be considered part of the *growing season*. It is not hard to see what the implications of this are if you realize that wherever there is a growing season of at least 5 months' duration, people can grow crops. Where it is shorter than this,

Cherrapunji, India, has the record for the world's greatest monthly rainfall (9300 mm in July 1861); the greatest annual rainfall (26 461 mm during August 1860 to July 1861); and the greatest average annual rainfall (11 633 mm).

Japan Information Centre
Consulate General of Japan, Toronto

High precipitation usually results in thick forest growth. In this highland reforestation project in Japan, three stages of tree growth may be seen: seedlings in the foreground, saplings in the middle ground, and mature trees, ready to be harvested, in the background.

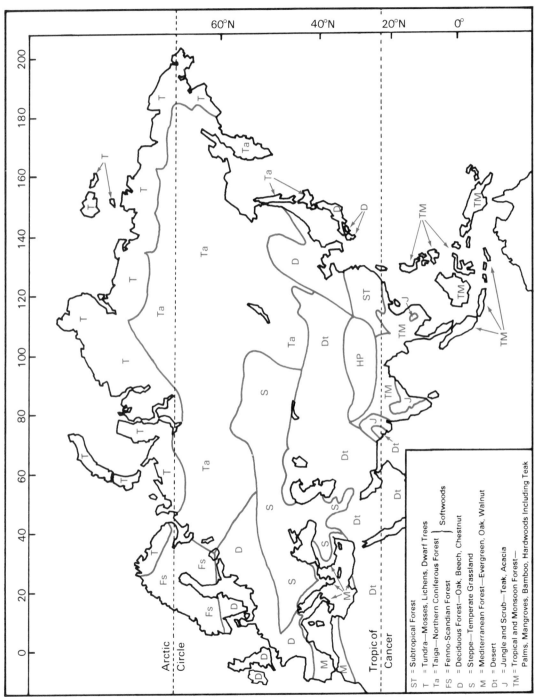

Figure 1-5. Eurasia — Natural Vegetation

ST = Subtropical Forest
T = Tundra—Mosses, Lichens, Dwarf Trees
Ta = Taiga—Northern Coniferous Forest ⎫
FS = Fenno-Scandian Forest   ⎬ Softwoods
                                                              ⎭
D = Deciduous Forest—Oak, Beech, Chestnut
S = Steppe—Temperate Grassland
M = Mediterranean Forest—Evergreen, Oak, Walnut
Dt = Desert
J = Jungle and Scrub—Teak, Acacia
TM = Tropical and Monsoon Forest—
        Palms, Mangroves, Bamboo, Hardwoods Including Teak

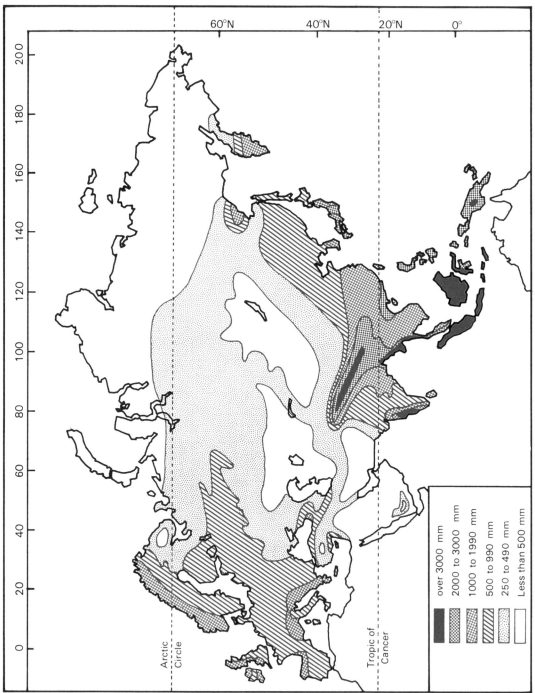

Figure 1-6. Eurasia — Total Annual Precipitation in Rainfall Equivalent

over 3000 mm

2000 to 3000 mm

1000 to 1990 mm

500 to 990 mm

250 to 490 mm

Less than 500 mm

there is a risk that if winter comes early, frost can kill just-ripened crops. Or, of course, winter might be late in leaving and so delay the time for planting seed. Either way, there is not enough of a safety margin to justify the heavy expenditure of time, energy, and capital (investment) used to clear the land, employ farmers, transport their crops to market, and keep the farmers themselves supplied with the equipment, fertilizer, spare parts, power, home comforts, and the 1001 things that they need to operate effectively. Of course, we must realize, too, that a long growing season should also go along with a sufficient supply of moisture and a suitable type of soil. While some such deficiencies can be overcome with irrigation and fertilizer, these are added costs that reduce profits. And it is profits that induce people to start farming in the first place!

Examination of Figure 1-7 (a) and (b) shows that, although growing season is important when one is considering where plants grow best, the *excess of temperature* above 6°C is important in determining the type of plant that can be grown. The amount of precipitation is important, too. Temperature, unfortunately, is a factor we cannot do very much about as yet. Perhaps we never should, because tampering with any climatic pattern is dangerous. Yet this is precisely what many people feel we have done already *unintentionally*. Pollution, and in particular dust particles in the atmosphere, put there in great quantities by high-flying aircraft, industrial smoke-stacks, soil erosion in the tropics, etc., seems to have acted as a barrier to the sun's heat, so that the earth has been slowly but steadily cooling since the end of the Second World War in 1945.

"For the past 25 to 30 years the Earth has been getting progressively cooler again. Around 1960 the cooling was particularly sharp. And there is by now widespread evidence of a corresponding reverse in the ranges of birds and fish and the success of crops and forest trees near the poleward and altitudinal limits."

"The decline of prevailing temperatures since about 1945 appears to be the longest-continued downward trend since temperature records began."

"With the rising population of Soviet Central Asia and increasing industrial need for water as well, the authorities have been obliged to consider diverting water from

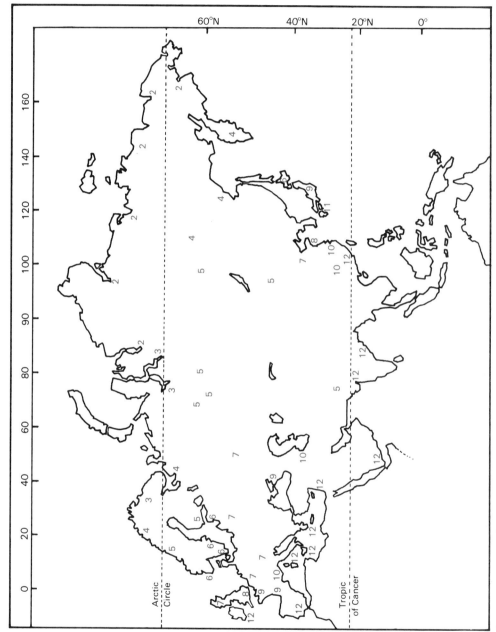

Figure 1-7 (a). The Approximate Duration of the Growing Season in Months

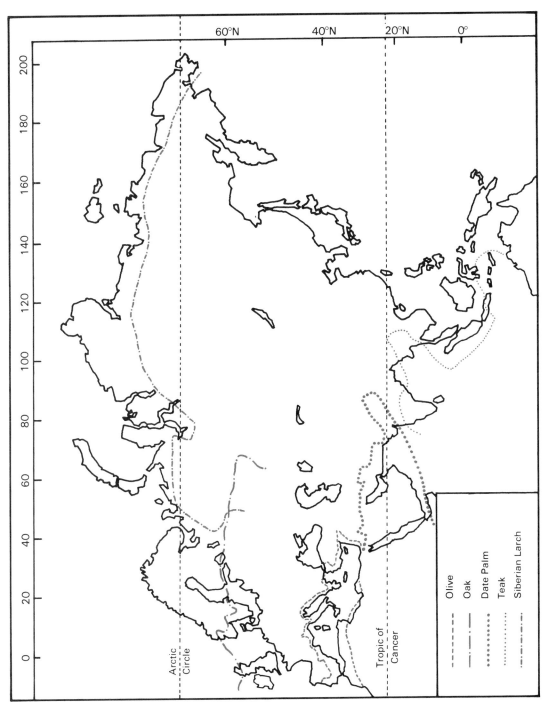

Figure 1-7 (b). The Growth Limits of Some Economically Important Eurasian Trees

the great rivers of Siberia which flow north to the Arctic Circle . . . if that ocean were converted into a normal saltwater ocean free of ice (it is the fresh water from the Siberian rivers which lies largely upon the surface of the Arctic Ocean and it is this water which freezes into a layer of continuous ice), most of the Arctic would be on an average 10°C to 20°C warmer than it is now, and over 30°C warmer in the winter time. This could affect all the climates of the northern hemisphere."
(reprinted from the *Unesco Courier*, August/September 1973.)

Changes in climate in Eurasia are nothing new, of course. In some parts of the land-mass, it is a mere 6000 years since the glaciers of the last great Ice Age retreated. In the extreme north or at high altitudes, there are still ice caps and glaciers that are the shrunken remains of the great continental ice sheets. And now the climate is getting colder again.

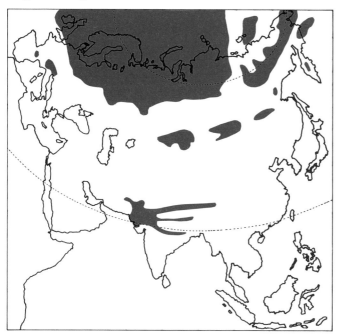

Figure 1-8.   The Maximum Extent of the Ice Cover During the Last Great Ice Age

The evidence for the recent Ice Age is overwhelming. Altogether, the ice covered 10 250 000 km². It gouged and eroded many lands to the north, depositing the debris as a blanket over the lands to the south. Fiords, deep

Swiss National Tourist Office
The summit of the Alps — Mont Rossa, 4165 m. During the last Ice Age, most European hill and mountain ranges looked like this: covered with a mass of snow and slow-moving glaciers. Only a few mountain peaks can be seen.

mountain valleys, and lake hollows are complemented by plains of boulder clay and till, moraines, and fertile *loess*, a wind-blown dust that was picked up by air currents at the edge of the areas of glacial erosion.

Imagine if you can the effects of another Ice Age. Much of central Europe and Asia would be covered by ice, and the land at the fringes would be Arctic in climatic type. Seas would be lower and planetary winds would have changed. Familiar plants and grasses would no longer be seen in their old haunts. Cities would have to be abandoned. The Sahara Desert would return to the green life it once knew as the European climatic belt is pushed south. Arabia would become *Arabia Felix* of the ancients, a name they bestowed on the land when it was much more verdant than today. Mesopotamia (Iraq today) would regain its former agricultural glory, as would the long-abandoned valley civilizations of the Indus River in Pakistan.

Who knows? While some meteorologists and climatologists preach caution, others are quite sure that all known civilization has sprung up merely in a warm interlude in what is predominantly a continuous Ice Age, and that much of the northern hemisphere must return to icy conditions in a few thousand years more.

The climate of Eurasia is a fascinating but complicated topic. There are large gaps in our knowledge of its rhythms. It varies across the face of the gigantic landmass and changes through time in a way that is hard to perceive because our life-times are so short, the changes are so gradual, and weather data have only been kept continuously for a century or so. In many areas, data are still not recorded.

What is required is a lot more research, money, time, and effort to solve old weather mysteries and discover new ones. Governments must co-operate more and expand the efforts of the World Weather Watch and similar activities. Satellite scanning and computer analysis play an increasing role in helping the WWW (an offshoot of the United Nations) predict weather for long periods ahead. Already it tracks cyclones and storms and monitors weather everywhere in the world, issuing warnings, advice, and alternatives. But you must do your bit. As a world citizen, you have a duty to understand as much as you can about world weather and climate. Changes anywhere have an effect on you.

"Thunder before the beginning of spring 49 days of bad weather will bring." (China)

Inscription on a stele of the Han period (206 B.C. to A.D. 220). Reprinted courtesy UNESCO Courier

QUESTIONS AND EXERCISES

1. Find out more about meteorpsychiatry and the effects of weather and climate on people.

2.

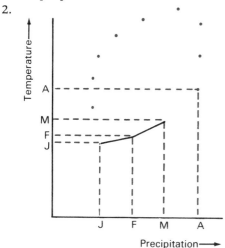

A *hythergraph*: each month is a point on the graph, and the 12 points are connected with lines as shown. Intercept lines (dotted) are omitted.

Construct *one* hythergraph suitable for plotting a shape for one example of each type of climate in the text. Be sure that your precipitation and temperature scales are large enough to allow all your shapes to go on one graph. That is most important. You might have to glue or tape several sheets of graph paper together to arrive at a sheet large enough for the shapes to be plotted out clearly.

Before you actually start to plot the shapes, label the parts of the hythergraph that correspond with the following terms: *hot and wet, hot and dry, cold and dry, cold and wet.*

When you have done all the preliminary work, decide on a different colour for each climate shape. *Then* begin!

When you have finished your climate graph shapes, name each city directly on the graph, but use a *legend* to say what type of climate the colour of the line means.

3. Use your library, friends, parents, *anybody* to find out as much as you can about other weather sayings that are used to predict rain, sunshine, wind, frost, cloud, etc. Try and give an example of each. Perhaps you can even say what their scientific basis is.

4. Trace a map from Figure 1-7(a), which shows duration of the growing season. On *your* map, try to draw lines at intervals of one month.

Compare your map with an atlas map that shows the distribution of population in Eurasia. Write down any conclusions that you have about the 2 maps.

Is there any area on the growing-season map where it looks as though many more people could live than it appears actually do from the population map?

Read the text again if you have forgotten what length of growing season is required for agriculture.

If you did find an area that looks as though it could be farmed now but does not appear to be, try to state what the drawbacks are to agriculture there. You will have to consider such things as rainfall, transportation, distance from markets, maybe other things, too, that strike you as being important.

5. What growing season line best separates *deciduous* from *coniferous* forest? Why is it that coniferous trees are better adapted to life north of that line?

Why is it that little of the *original deciduous forest* cover now remains in Eurasia? What happened to it? Is that a good reason to explain why little *steppe* remains either?

2. POPULATION DISTRIBUTION

GLOBAL VIEW

Approximately 75% of all the people in the world live in Eurasia. In the mid-1970s this means that out of about 4 billion people in the world, 3 billion or so live in Eurasia. If it's around 1980 when you're reading this, then it means that out of at least 4.5 billion people in the world, about 3.4 billion are living in Eurasia.

If you are like most people, the concept of *billions* doesn't mean very much to you. Maybe you have seen the display of a million dots at the Ontario Science Centre in Toronto; if you have, then try to imagine a thousand of those displays! Or perhaps you can look at it this way: you probably eat three meals a day for 365 days a year, possibly skipping the occasional breakfast or lunch, so let's say you eat an average of 1000 meals a year. In order to eat a billion meals, how old would you have to live to be? Or think of the second hand on your watch — it will have to keep ticking for over 30 years in order to count off a billion seconds!

You can see, then, that the world certainly contains a lot of people. In fact, for every Canadian there are approximately 175 *other* people in the world, and about 130 of *them* live in Europe and Asia. Figure 2-1 shows that they are not evenly spread all over Eurasia. There are, in

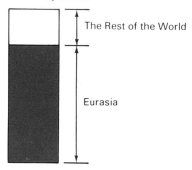

World Population

The Rest of the World

Eurasia

If each walking pace you took averaged 50 cm, how far would you have to walk in order to take a billion paces?

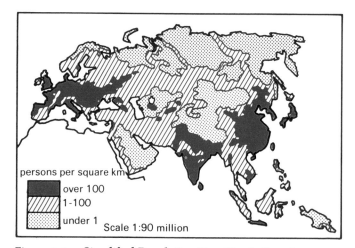

persons per square km

- over 100
- 1-100
- under 1

Scale 1:90 million

Figure 2-1. Simplified Population Densities in Eurasia

fact, 3 distinct major population clusters, and a lot of empty space in between. The 3 important areas of population are:

1 Europe, especially the lands surrounding the southern part of the North Sea, where the Netherlands, for example, has a population density of nearly 400 people per square kilometre.

2 The Indian subcontinent, especially the lands of the Ganges Valley, where Bangladesh, for example, has a population density of well over 500 people per square kilometre.

3 The Orient, especially eastern China and southern Japan. Taiwan is quite typical of this cluster, with a population density of about 400 people per square kilometre.

In contrast to these 3 highly populated areas are 3 very slightly populated areas: northern Eurasia, from Scandinavia across to Siberia; central Asia, from the Caspian Sea across to western China; and the Middle East, from Iran to the Red Sea.

One of the things that we, as geographers, have to do is try to explain why these great contrasts occur. Why do so many people live in the crowded areas, and why do so many avoid the ones that are uncrowded? The basic answer, for people *en masse* and over a long period of time, is that they want to. But it is not enough to say that; we must try to examine *why* they want to.

So let's start. First of all we shall investigate the possible reasons why so many people live in Europe.

Novosti Press Agency

Samarkand, in the thinly populated desert of central Asia.

HIGH POPULATION DENSITY IN EUROPE

Farming is fairly easy to practise over most of Europe, and it laid the foundations for a large and crowded population. Farming itself is favoured by numerous areas of fertile soil, chiefly river silts deposited in hundreds of river valleys, but also including loess and clay of glacial origin. These deposits usually occur in lowland areas, and if you look at an atlas map of Europe you will notice that most of Europe's lowland areas are in the lands around the southern shores of the North Sea. Lowlands also offer the advantage of milder temperatures (generally, temperatures decrease with height at a rate of about 6°C for every 1000 m), and this too favours farming. (Why do you think this should be?)

Loess is a fine, sandy dust that binds into a very fertile soil. It probably came originally from deserts. It is now to be found in broad, deep patches all the way from northern France (where it is called *limon*) to northern China. Some of the largest deposits of loess in Europe exist in the North German Plain, especially in the Cologne region and the area between Hanover and Leipzig.

Farming is also aided by reliable supplies of water. For most days in the year the most common winds (called *prevailing* winds) are from the Atlantic Ocean. The prevailing Westerly Winds carry moisture from the Atlantic Ocean and deposit it as precipitation over the lands of Europe. Most of the precipitation actually falls in the mountains rather than in the lowlands, but it drains out of the mountains as rivers, and so it is available for use in the lowlands. (Of course, the rivers also bring down fresh supplies of silt from the mountains.) The Westerly Winds also help farming in another way; they keep the winter temperatures at an unusually high level. Normally at these latitudes (What latitudes? Check in your atlas), the winter temperatures can be very low. But because of the Westerly Winds, Europe is kept mild in winter. The reason is the Gulf Stream Drift. Warm water emerges from the Gulf of Mexico throughout the year, and it is eventually blown across the Atlantic by the prevailing Westerly Winds. In time it reaches Europe. This warm Gulf Stream water does not make much difference to Europe's temperatures in summer, because the sun then is high in the sky and Europe is warm enough anyway. But in winter, when the sun is low in the sky, the warm Gulf Stream Drift is really useful to have around. Not only does it reduce the risk of frost in the fall, but it also enables farmers to get off to an early start in spring.

But farming is not the only reason for Europe's high population density. Industry is also important. The Industrial Revolution began in Europe, and since 1800 or so Europe's people have relied heavily on industry for jobs. The advantages for industry are many. Chief among them is the availability of reliable water power. In fact, the very earliest industries depended on water power. You can see, then, how the Westerly Winds were of help to industry as well as farming. Reliable precipitation formed many rivers that rarely froze in winter. Today, of course, direct water power has been replaced by hydro-electric power—which also requires a reliable water supply. Europe also has a great number of separate coalfields. Many of them are rich and easy to mine, and even today new ones are being discovered. (For example, a new coalfield was discovered in eastern England in 1974.)

Good transportation systems are also necessary to industry, and here too the regular precipitation helps. The numerous large rivers flowing through the lowlands have been used for hundreds of years for cheap transportation.

Prevailing winds are simply those which blow most often. There may be winds from other directions as well, and sometimes these other winds may be very strong. But only those winds that blow most often are called prevailing winds.

A simple *wind-rose* showing prevailing south-west winds.

Typical January mean temperatures along the 50°N latitude line, from west to east, are:

France	5°C
West Germany	0°C
East Germany	−5°C
West U.S.S.R.	−10°C
Central Siberia	−20°C
Eastern Siberia	−30°C

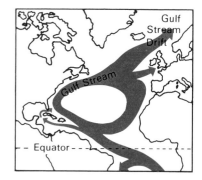

Chief among them is the Rhine. And because of the warmth from the Gulf Stream Drift, the waterways can be used almost throughout the year. Compare these European rivers with the St. Lawrence. It used to be closed for 4 months every year until special ice-breakers and heated locks were used to keep navigation going. In Europe, navigation is much easier. The lowlands of Europe also help the waterway system, because it is relatively easy to construct canals linking the separate rivers. Furthermore, it is fairly easy to construct roads and railways on these flat lands.

Trade has always been important to Europe. Most of the great voyages of exploration sailed from Europe, opening up trade with other lands. A major advantage to trade has been Europe's many natural seaports; the coasts of Europe are very broken into peninsulas and islands, and there are lots of bays and river mouths. These natural ports are ice-free (the Gulf Stream Drift again) and generally deep-watered. Over the years the development of trade has led to other commercial activities, such as banking and insurance. These *service industries* also help to support many people in Europe, thereby contributing to the high population density.

In summary we can identify 4 main factors that help to create a high population density in Europe:

1   farming, which is varied and generally efficient, characterized both by very high yields per hectare and by very high yields per unit of labour input;
2   industry, which was pioneered in Europe in the 1700s and 1800s and is now both varied and efficient;
3   trade, which is not only an old tradition but now also a very valuable modern practice;
4   service industries, which create jobs not only in banking and insurance but also in the care of tourists and in general transportation.

HIGH POPULATION DENSITY IN THE INDIAN SUBCONTINENT

On the whole, India does not have so many more people per square kilometre than do most of the countries of Europe. Indeed, it has less than some; India has 168 people, Italy 180, U.K. 236, West Germany 241, Belgium 302, and the Netherlands nearly 400 per square kilometre.

Nevertheless, most people have the distinct impression

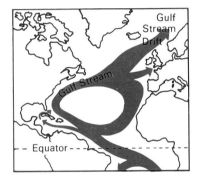

Lloyd's of London is one of the world's largest insurance houses. It specializes in transportation insurance, particularly shipping. Most countries of the world have their ships registered at Lloyd's. The business began in 1689 when Edward Lloyd's coffee house in London's Tower Street became the chief meeting place for a group of businessmen who were interested in shipping ventures.

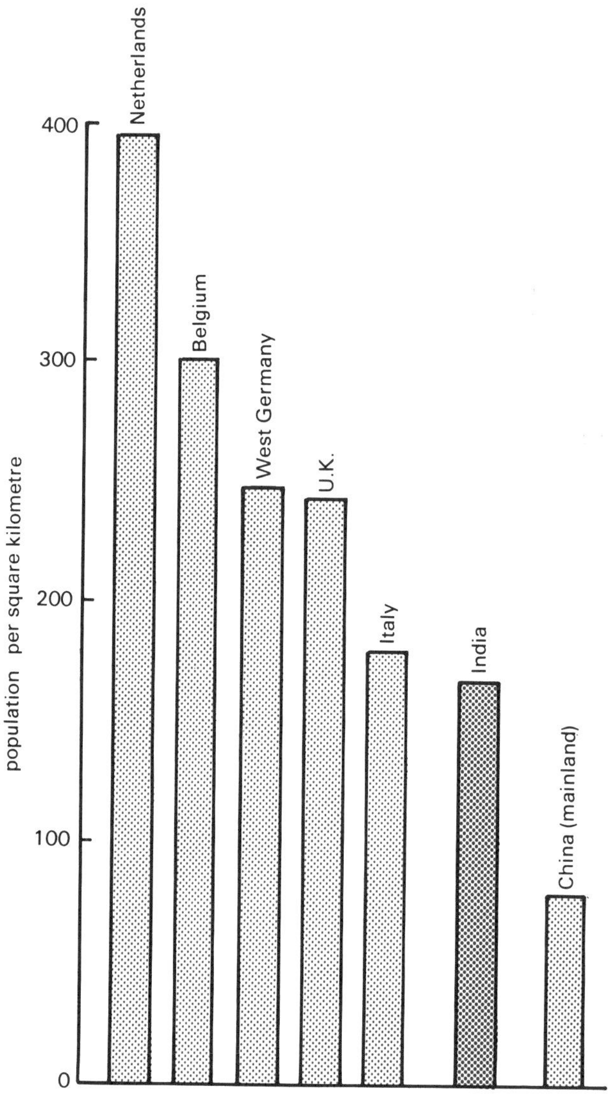

Figure 2-2. Some Comparative Population Densities

that India is the most crowded place on earth. Why do you think this is so? One of the most important reasons is that most of the people in India are very poor. The housing creates a visual appearance of crowding, because the small houses of the poor are very close together and close to the street. Since temperatures in the sub-continent often rise above 40°C, people usually do not want to stay for long indoors. It would be unbearably hot and airless. So the general movement of people to the outdoors also helps to create the impression of crowding.

Naturally, there is more to the story than that. For one

thing, India has been settled by people for thousands of years. During that time, naturally, there has been plenty of opportunity for a large population to build up. Part of the reason for this large population is the environment, especially the climate — of the monsoon type — which is favourable. Now that does not mean that everything about the climate is perfect; in fact, it is often too hot for the people to bear. What it does mean is that *on balance* the climate offers enough advantages to make people want to live in the area. The advantages are twofold: First, the temperatures are high enough throughout the year to permit a wide variety of crops to be grown, and indeed in some places even 2 crops a year. Second, the monsoon rains come in the summer when the need for water for crops is greatest. The combination of great heat and heavy rain in the summer months favours the growing of rice, and this is another advantage. Rice produces more Calories of foodstuff per hectare than either wheat or corn, and so it can support more people. It is not just by chance that the maps of the chief rice-growing areas and the highest population densities are so similar; nor that the map of the main rainfall areas is similar to both.

The great heat and heavy seasonal rain of the Indian monsoon has also presented problems. Insect-borne diseases are especially common, often making many of the people listless. Drought is a bad problem too, of course, because a climate with seasonal rain by definition also has seasonal drought. The monsoon rains usually fall from June to November. Thus from November to June there is drought — and that's a long time. Many people starve. The combination of frequent disease and seasonal starvation has not promoted economic growth. Other factors have also helped to slow down economic development. For example, the ancient religious and social traditions that people cling to may also be partly responsible. Whatever the reasons, there is no doubt that India, Pakistan, and Bangladesh all have an extremely slow rate of economic growth. The average income per person is among the lowest in the world.

Farming, which is the mainstay of existence in the sub-continent, shows all the features of lack of development: low yields per hectare; low yields per unit of labour input; little use of fertilizer; little use of machinery; little use of pesticides; poor storage facilities; poor marketing facilities (if any); and almost no knowledge of more recent scientific advances in such areas as seed selection, cattle

Karma is the Hindu word for fate. Every Hindu believes that he has his own *karma*, which he cannot escape. Thus a Hindu passively accepts the things that happen to him, however bad they may be. He believes that the more willingly he accepts his present life then the better will be his life after he is re-born. Belief in reincarnation is therefore a consolation which makes starvation easier to bear.

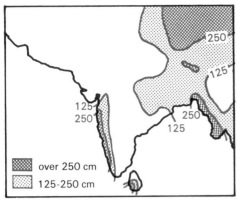

Annual Rainfall in the Indian Sub-continent

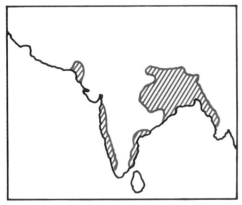

Major Rice-growing Areas
in the Indian Sub-continent

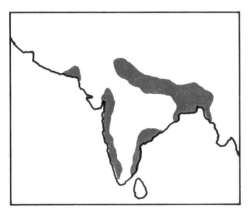

Chief Populated Areas in the
Indian Sub-continent (over 200/Km^2)

Figure 2-3. Some Salient Facts about the Indian
Sub-continent

breeding, and irrigation. Because the farming methods are poor, the farmers are poor; but because the farmers are poor, the farming methods remain poor. Some attempt has been made to improve things. The most promising hope lies in the so-called *Green Revolution*, which has introduced special high-yielding seeds to India. Unfortunately, the new seeds require a lot of fertilizer and a lot of irrigation, so that only the richer farmers can afford to use them. The rich farmers then buy out the poorer farmers, who have nowhere to go but the towns to look for work. Work is almost impossible to find. And so the towns become even more crowded (which is why so many people live *on* the streets), and starvation and poverty continue to characterize the population.

There is still one last reason why India is so crowded: the high rate of *natural increase*. Natural increase is the difference between the birth rate and the death rate. The larger it is, the more additional people there are each year. In India the natural increase rate is about 25 or 26 per 1000 per year, which means that for every 1000 people already in the population there are an extra 25 or 26 each year. You can easily figure out how many additional people that makes every year: start with a population of about 550 000 000. There are 550 000 thousands in 550 000 000, and for every one of those thousands there are, say, 26 extra people each year. So that makes a total addition of 14 300 000 people (550 000 × 26) each year. In 2 years that's more than Canada's entire population! One more point: the natural increase rate is, of course, *compounded*. Think about it. Pretty soon, India will be as crowded as the U.K. or West Germany.

In summary we can identify 3 factors helping to promote a high population density in the subcontinent:

1 Farming, which occupies about 90% of the total population and is characterized by low yields. Most of the farming is carried out on a self-sufficiency basis.
2 Poverty, which is partly the result of slow economic growth. Because food production is at a low level of technology, vast numbers of people can be supported only in poverty. If the poverty is to disappear, then so must the low-level technology; the two must both exist together, or both disappear together.
3 High rate of natural increase, which is adding millions of extra people yearly to an already tight situation. Whenever there is any increase in food output, there are also millions of extra mouths to be fed.

Indian Government Tourist Office
A peasant farmer of N.W. India.

Less than 20% of India's population lives in its towns. Even so, this is more than the combined populations of France and Britain. And within India's towns it is estimated that over 20% have recently moved in from the surrounding countryside.

HIGH POPULATION DENSITY IN THE ORIENT

The Orient consists of China, Taiwan, Korea, and Japan. If you were to argue strongly, then you might be able to include the Philippines, Vietnam, Laos, Thailand, and Khmer; but you would be getting yourself into the same sort of argument as if you tried to say that all Canadians were just like Americans. In the 4 countries of the Orient the basic causes of high population density have been farming and time. The lands have been settled and farmed for thousands of years. Each year the floods caused by the monsoon rains have brought down silt to replenish the fertility of the well-used fields. Rice has been the staple crop for most of the people, who have subsisted on a multitude of small farms for generations. Just as in India, crowding and poverty became typical. However, all 4 countries have escaped from the type of situation that prevails in India.

Japan started to move away from potential disaster in 1868, when a new emperor, Emperor Meiji, decided that Japan should develop itself in the manner of western countries. So Japan sent people out across the world to learn about foreign industry and trade. They learned, and went back to Japan. Nowadays, of course, Japan is the world's third most important economy, with a huge investment in manufacturing industry. Since the time of deliberate industrialization, Japan's population has increased from about 33 000 000 to about 105 000 000. So you can see very clearly that industry and trade help considerably to support a high population density. With the growth of trade and industry, Japan is now following the European pattern of continued growth into the service industries. Japanese banks are now among the world's leaders, and their shipping and airline services circle the globe.

Korea is following the same route as Japan did, but it started much later, in the mid-1950s. Much of Korea's industrial progress has been actively guided by Japan. After all, they are neighbours, and Korea was the obvious place for Japanese firms to go when they wanted to expand (and had no more room in Japan itself). Korea's economy still has a long way to go compared with Japan's, but it's on the way. So is Taiwan's, which is also following the same route.

China is a different matter. After being torn by inva-

John Bedford

In Hong Kong, many people live in houseboats to save space.

Sue Mason

Rice is a staple crop for most oriental people.

Japan National Tourist Organization

Japan is now the world's third most important economy, with a huge investment in manufacturing.

sions, rebellions, civil war, and revolution, China slowly began to put itself together in the 1950s. The communists, led by Mao Tse Tung, pursued a policy of economic growth based entirely on the concept of self-sufficiency (after some Russian help in the early days). This was quite a different way to economic growth than that followed by the other Oriental countries, which have relied considerably on trade. By means of strong political power, the Chinese have successfully managed a rising rate of economic growth. Farming is still the direct mainstay of 90% of the population, but technology is being ingeniously applied. As a result, poverty has been largely wiped out. If you can get films from your local library, try to obtain either *Red China* or *Mao's China*; they will give you a good insight into what life in China is like. As a contrast, see also if you can get *Asian Earth*, which deals with the life of a poor family in India.

In summary we can identify a mixture of factors that help to produce a high population density in the Orient:

1 Farming, which is being modernized very effectively, so that Japan is now one of the world's leaders in farming technology. The other countries are behind, but following.

2 Time, which has given the Chinese in particular a great opportunity to be the most numerous people on earth.

3 Industry, which is varied in efficiency, ranging from the Western techniques of the Japanese to the self-sufficient methods of the Chinese.

4 Trade, which characterizes the growth patterns of Japan, Korea, and Taiwan, but not China.

5 Service industries, which are at present only well developed in Japan.

AREAS OF LOW POPULATION DENSITY

All the areas of low population density suffer from drought, as a glance at Figure 1-6 will show. There is, in fact, a broad band of desert all the way through Eurasia, from the Arctic Ocean to the Red Sea. In the north, from Finland to eastern Siberia, precipitation is generally within the range 250 mm to 500 mm per year, which would just about be enough to grow a very limited variety of crops if only the temperatures were high enough. In central Asia, conditions become more like true desert: Kashgar

Japan National Tourist Organization

Japanese airline services now encircle the globe from their home base at Tokyo International Airport.

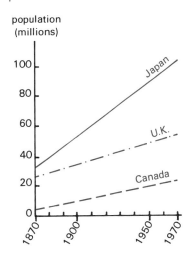

The Japanese keep a breed of beef cattle called Kobe. Kobe cattle are very specially cared for. Their keepers spray them with beer (Japanese beer is very good, too), and then massage the beer into the flesh of the animals. This is intended to make the flesh very even and non-fatty in consistency, so that when the animals are slaughtered the beef simply melts in your mouth. Kobe beef costs the earth, and most Japanese cannot afford it.

in western China has about 80 mm of precipitation per year, while Astrakhan on the shores of the Caspian Sea has about 150 mm. Farther south, towards the Red Sea, precipitation remains very low (Karachi 200 mm, Isfahan 100 mm, Kuwait 130 mm, Mecca 60 mm). Here, most of the landscape is true desert.

The shortage of water for farming, coupled with the extraordinary low temperatures and long winters of the far north, effectively limit people's use of these areas. Thus population densities are very low. But there are still *some* people; the lands are not quite empty. Mostly the people are wandering bands of herders, tending sheep, camels, and goats), moving from sparse pasture to sparse pasture and from water hole to water hole. They are collectively called *nomads*, people who wander. There are also scattered settlements at the water sources (springs or streams coming from the mountains), but most of these settlements are quite small. The chief exceptions are the newly industrialized towns of Siberia. Here the Russian government has been deliberately building up steel and engineering industries to make use of the huge water-power potential on some of the giant Siberian rivers. Irkutsk, Krasnoyarsk, and Novosibirsk are examples of these new Siberian towns. The former disadvantage of remoteness from existing population clusters has been somewhat offset by the construction of the Trans-Siberian Railway and by the development of air services (the internal U.S.S.R. airline, Aeroflot, is one of the largest in the world).

Yet another reason for much of the relative lack of crowding in interior Asia is the mountainous nature of the land. The Himalayas are the chief range, but they are only one of many. The mountains effectively limit the development of farming, as well as seriously hindering the growth of trade and industry.

In summary we can list the following factors as contributing to the low population density in northern Eurasia, central Asia, and southwestern Asia:

1   drought, ranging from semi-arid conditions in the north through to true desert in the centre and southwest;

2   low temperatures, especially in the north, where, for example, the town of Verkhoyansk has a regular January monthly mean temperature of $-50°C$;

3   isolation from existing population clusters, chiefly those of Europe;

Trains on the Trans-Siberian Railway are scheduled to make the 9300 km run from Moscow to Vladivostok in almost exactly 8 days. What is their average speed in km/h?

4   mountains, which acted directly to limit the amount of suitable land for farming and indirectly to increase the isolation already imposed by distance. Thus the people of central Asia were doubly remote from the population clusters in China and India. This is one of the reasons why Siberia is governed from Moscow rather than from Peking or New Delhi.

POPULATION GROWTH

If you could cut the shape of Eurasia out of a piece of card (which you can't really do because of all the coastal islands), then balance it on a needle, you would probably find that the balancing point was somewhere around Novosibirsk in the U.S.S.R. Now, that would just represent the land. If you could weight the cut-out with people, you would find that the balancing point shifted towards western China. That is of course because there are lots more people in China and India (and Japan and Indonesia) than there are in Europe.

As time goes on, the balancing point is moving closer and closer towards southeast Asia. More and more people are being added to the already large populations in India and the Orient without any equivalent large number being added to the large population of Europe. And so the population balance of Eurasia is continuously shifting.

We have already noted the growth of population in India. We should add that similar increases are taking place in Bangladesh and Pakistan, as well as in all the small countries of southeast Asia. For example, the natural increase rates in Bangladesh (32 per 1000 per year) and Pakistan (30) are accompanied by similar rates in Taiwan (23), Malaysia (35), Sri Lanka (22), Thailand (33), and Korea (28). The Japanese rate was also high at one time, but it has now declined to about 12. In China the rate is on the way down, though it is still high, at about 18. India's, on the other hand, is on the way up. Indeed, the United Nations forecasts that India's population will be larger than China's by the early 2000s, when both are expected to be over a billion each.

One of the most direct effects of this growth of population is that population densities will inevitably increase. For example, in about 25 years, when India's population reaches one billion, the population density in India as a whole will be over 300 people per square kilometre, compared with only about 170 people per square kilometre

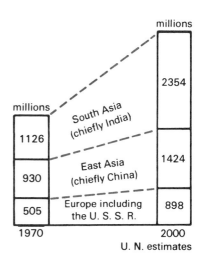

now. Since most of those additional people will live where the present ones do — in the rice-growing areas of the south and northeast — you can imagine the sort of crowding there will be then.

In Europe, on the other hand, population growth, which was rapid at one time, has slowed down almost to a standstill. For example, West Germany has a natural increase rate of 2 per 1000 per year, East Germany 0, France 6, Italy 7, the U.K. 4, Hungary 3, and so on. As a result of this slow growth, population densities are not likely to change drastically in the future. But Asia, especially south and southeast Asia, is altogether a different matter. Look for trouble in these areas as you get older. Whenever more and more people are forced to share the same amount of space, there is more likely to be friction, perhaps even riots, possibly revolution.

QUESTIONS AND EXERCISES

1. Make up a checklist of factors that influence the distribution of population. You should have 2 sections: one of factors that favour population, and another of those that discourage settlement. How does your checklist apply to your own area?
2. Draw a divided circle graph to illustrate the proportion of the world's population living in Eurasia.
3. Explain what is meant by these different forms of irrigation:

flood irrigation	perennial irrigation
sprinkler irrigation	shaduf irrigation
gravity irrigation	pumped irrigation
qanat irrigation	sakia irrigation

4. Write an account of the different ways in which the Gulf Stream Drift influences Europe.
5. What are the advantages and disadvantages of the monsoon system of rainfall?

3. TRANSPORTATION

SPATIAL INTERACTION

Louis Brun glanced at the glowing lights of the control panel. Everything was just as it should be. The giant plane was approaching the runway at Peking Airport, and Louis was looking forward to spending a few days flying "demos" around northern China and the western Pacific. The Anglo-French team that had designed the supersonic passenger plane, and the government agencies that had organized the sales mission, would be proud of the way the plane had behaved on the run from Paris. They had left Paris at 20 00 h and had taken only a little over 4 hours to reach Peking. It was now shortly after 08 00, and the sun was shining brightly. Like a gigantic bird, the plane came down with its wings tilted backwards. The wing wheels touched down with a squeal and a puff of smoke. Gradually the droop-nose came down and the nose-wheel settled on the runway. It was a perfect landing. Louis smiled.

Supersonic means faster than sound. Sound waves travel through the air at a speed of about 1150 km/h; anything that travels faster than this is said to be supersonic. Aircraft that fly at the speed of sound (and the actual speed varies slightly according to the altitude of the plane, because of differences in air pressure) are said to fly at *mach 1*. An aircraft flying at twice the speed of sound is said to have a speed of mach 2, and so on. The Anglo-French *Concorde* has a speed of about mach 1.5. It is the only supersonic passenger aircraft in the Western world, the Americans having cancelled construction of theirs back in 1970. The Russians also have one.

Two thousand kilometres farther east, Kendo Bengoshi stood in line on the suburban Tokyo station platform, waiting for the Tokaido Express. He knew he would not have to wait very long; the trains ran every 10 min. His seat was reserved, just like all the others on the train. Eventually the train came, and the waiting passengers filed aboard. Within a minute the train was moving again. Kendo opened his briefcase. He had just over 3 h before the train arrived at Osaka, over 500 km away, to go over

35

The so-called bullet trains of the Tokaido Line were built especially to serve the densely populated region between Tokyo and Osaka, but the line has since been extended to north and south. The track is used only by the bullet trains, so there are no slower trains to impede the progress of the bullets. The track was also designed so as to eliminate all sharp curves and steep grades, thereby permitting maximum operating speeds (up to 200 km/h in places). Running every 10 minutes, the trains are a mixture of express and local-stopping. Most of them are local-stopping trains, but because of the tight operating schedules the stops are rarely longer than a minute at any one station. Control is provided by a computerized Automatic Train Service, which will automatically brake trains if anything goes wrong.

the contracts he had prepared for his firm's sales of ships to a Middle eastern country.

Meanwhile, Gulam Vatsava was just going to work in the hiring hall at Bombay docks. The harbour was full of ships, tied up and quiet, with only the spicy smell of breakfast cooking and the indefinable mixture of salt and metal smells to make Gulam glad he was alive. Gulam wanted very much to be happy. He had not managed to get work for several weeks, and he was desperate. First of all, back in May, his youngest child had died, and his wife had been very sick. Food was scarce, and there had even been rioting in the streets near the government food stores. After that had come the monsoon floods and severe disruption of road and rail traffic. Then, too, there were strikes on the railways. During June, and into July, the ships had stayed away from Bombay; there was little for them to call for. In late July, the economic recession which had gripped much of Europe for many months seemed to ease slightly. There had been new elections in Italy and the U.K., and new strong governments had been elected. People there began to feel more optimistic, and businessmen began to invest in factory expansions once more. Gulam did not know any of this. He merely went to the hiring hall looking for work. It was somewhere to go. But because of the renewed expansion of economic activity in Europe, there were more ships in the harbour. Gulam noticed that. He also knew that the road and rail routes were open once again in the lands behind Bombay; the monsoon floods were ended, and the strikes were over. As it turned out, he was hired this day — loading a

Mick Horner

Dislocation of road traffic caused by monsoon floods in India.

freighter with raw cotton to take to Manchester in England. He was glad he was alive.

In the giant Tyumen oilfield in western Siberia, 5000 km to the north, Sergei Vasiliev was just finishing his night shift. Around the plant he walked, checking that the pumping operations were running smoothly. They were. Sergei nodded to himself; his job was, he felt, rather lonely — just looking after this great automated plant. However, it paid well, and it was easy work. He initialled the report log and went back to his inner office. The oil continued to be pumped. Sergei knew that most of it went to Moscow, over 1000 km to the west. He didn't know that the Japanese were trying to persuade officials in Moscow that some of it should also be piped eastwards to the shores of the Sea of Japan, where the Japanese could buy it. The talks would continue for a long time. Meanwhile, Sergei signed out as his replacement came in. Then he went home to breakfast.

These incidents show different types of spatial interaction. On some occasions the interaction is at an official level — government to government — as was shown by the flight of the supersonic passenger plane; on other occasions it is strictly a business matter, such as Kendo's arranging the sale of ships to the Middle East. At other times, spatial interaction may occur without the people who are affected knowing very much about it. Gulam, for example, was partly affected by what happened in Europe, but he was not aware of this. Nor was Sergei aware of possible changes that might occur in his job. Yet in all cases, what happened in one place was to some degree influenced, or likely to be influenced, by what happened in some other place. This sort of influence that one place exerts on another is called *spatial interaction*. Mostly it relies for its effectiveness on adequate transportation, because if places are remote and inaccessible, it is not likely that other places can readily influence them, so there is unlikely to be much spatial interaction. Transportation is, therefore, the key to spatial interaction.

The chief media of transportation are river, canal, sea, rail, road, air, and pipe. Each of these media (except pipe) can be used in a great variety of ways. Roads, for example, are used by people carrying loads on their backs, pack animals, animals pulling carts or wagons, pedal cycles, cycle-carts, and motorized vehicles of all sorts. These are all different conveyances. Similarly, rivers are used by

Cotton is grown extensively on the black volcanic soils of the *Deccan* region inland from the hills behind Bombay. Most of it is now manufactured into cotton goods in factories in the Bombay area, but some (about 20%) is still exported in its raw state, mostly to England.

Pack camels in Isfahan, Iran. *Mick Horner*

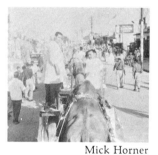

Swiss National Tourist Office

Pack mules in Switzerland.

Mick Horner

Bullock carts, Pakistan.

Sue Mason

Cycle-carts, Peking, China.

Mick Horner

Bus in North India.

Bicycles in Holland.

Netherlands Tourist Board

Mick Horner

A truck in Turkey.

Expressway in Switzerland.

Swiss National Tourist Office

canoes, barges, rafts, and even ocean-going ships. In the case of pipe, the medium itself is also the conveyance.

In addition to the variety of conveyances, there is also a great variety of things carried. For example, if we select just one medium — rail, say — and one type of conveyance — freight trains — then we are still faced with an enormous number of kinds of things carried — everything from pianos to explosives. And to top it all, the different media all connect many different places, and they are used in most places by a great variety of conveyances, collectively carrying a bewildering assortment of actual commodities. So where do we start in studying transportation?

Let's start by deciding what to leave out: we shall omit all the transportation systems covering local areas only, and we shall omit all the systems that are little used, even though they may be important to the people who use them. This means that we shall leave out local systems such as city rapid-transit systems and rural canal systems, wherever they exist in Europe or Asia. It also means that we shall miss the caravan trails across central Asia, the navigation networks in Indonesia, and much, much more. Nevertheless, we still have a lot left; so let's deal with it bit by bit.

London, England's rapid transit network includes about 500 km of subway track and about 7000 buses, mostly double-decked. It carries over 1 million people per day, and it loses money regularly. Tokyo's subway system is less extensive, stretching for only about 250 km, but there are few buses in Tokyo, and the subway trains are therefore crowded.

RIVER TRANSPORTATION

Figure 3-1 shows some of the major navigable rivers of Eurasia. However, don't think that navigation is always available on these rivers, or that it is equally available on all of them, either. Many of the rivers suffer from serious hindrances to navigation for at least part of the year. The great rivers of Siberia, for example, all freeze during the long winters. Even when the spring thaw comes there is still trouble, because the thaw melts the waters in the southern reaches first. Since all the rivers flow northwards, the mouths are still blocked by ice, even when the southern parts are flowing. You can see what this means — floods! So the use of these rivers for navigation is seasonal only. (You might also be aware, though, that the frozen rivers are used for truck transport in winter, so that they are not entirely useless as a transportation medium even when they cannot be navigated by ship.)

The monsoon rivers present other problems, because they almost dry up during the dry season. (When is that?) When the rains come, they usually do so abruptly and

Figure 3-1. The Chief Navigable Rivers of Eurasia

violently, so there tend to be a lot of heavy floods. As time passes, the monsoon countries are attempting to build dams to control the waters, so that run-off is delayed during the rainy season (thereby limiting floods) and made available instead during the dry season (thereby lengthening the time of navigability). The Mekong is an example of a river currently undergoing this type of development.

The rivers of western Europe are generally the best suited for navigation, because of the regular nature of the rainfall and the mildness of the winters. But even the European rivers are not perfect. They have periods of low water, when the precipitation comes as snow instead of rain, and they have floods, when the snows melt. Nevertheless, the disadvantages are not so marked as in Asia, and generally the rivers are well used. The Rhine is the classic example.

Marilyn Olsen

Barge traffic on the Danube at Bratislava.

Duisburg, at the confluence of the Ruhr with the Rhine, is the world's busiest river port.

CANAL TRANSPORTATION

Some of the canals shown in Figure 3-2 will take only small barges (as do the canals of England), but others will take very large barges (as do the canals of Europe, which will accept barges up to 6000 t). Canals are most evident in Europe. Here the geography — extensive lowland areas, regular precipitation, low evaporation rates, many rivers that could easily be connected by canals — and a mixture of early-industrialized nations combined to favour canal building. The canal-building period of European history was essentially the early industrial period too, when the cheap movement of bulky goods became an important consideration. This is still an important reason for the continued use of canals in Europe, of course, and it is also the reason why canals have not yet become so important in south and east Asia. In east Asia, indeed, the chief canal has nothing to do with industrialization at all; the Grand Canal of eastern China was constructed hundreds of years ago to make it easy to ship rice (in bulk and at low cost) from the rice surplus areas of the south to the rice-deficient areas of the north. But apart from the Grand Canal, there are no canals of much importance in south or east Asia. Industrialization has yet to make its demands over most of the area, and where it already has, in Japan, there is not enough lowland to allow canal construction. Japan has found it far better to use the sea for cheap bulk

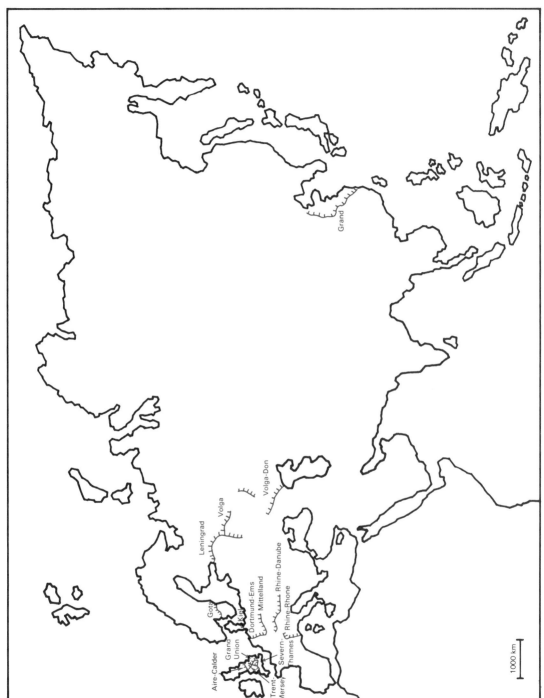

Figure 3-2. Some Important Canals

transportation, and the industrial towns are therefore located next to the sea instead of on rivers or canals.

Within Europe, however, canals are very important links between producing and consuming areas, especially when the products are heavy (iron ore, coal, steel, chemicals) and bulky (wheat, textile fibres, construction materials). The slowness of canal transportation generally makes canals useless for goods that are already made and just waiting to be sold. Nobody would send cars by canal, for instance, because the sooner they can be sold the better it is for the manufacturer. The other materials that we have mentioned — iron ore, etc. — are not quite the same, however, because they are used in much more of a continuous-process sort of way, and so long as there are continuous supplies at the using end it doesn't really matter very much how long they took to get there. For ex-

Bert Witvoet

A ship canal in Northern Holland.

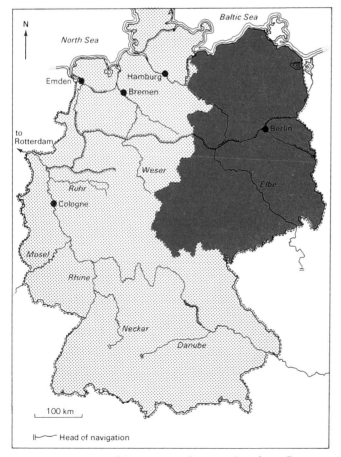

Figure 3-3.   Navigable Rivers and Major Canals in Germany

ample, the steel industries of the Ruhr receive their iron by barge; the iron takes weeks to reach its destination, but so long as the steel works have iron ore always available, it is not important how long it takes to arrive.

Canals not only offer cheap transportation for heavy and bulky goods, they also offer an alternative to road and rail transport. If canals could not be built so easily, the heavy and bulky freight would have to go by road or rail, and that would cost more. Canals therefore benefit European industry.

Where the canals become too small — or more exactly, when the barges become too big to fit into the older canals — then the disused canals can be developed to accommodate pleasure boating or fishing. Many of the English canals are like this.

Average costs of shipping bulk freight by canal barge are only about 20% of the costs of shipping similar freight by road or rail.

SEA TRANSPORTATION

As you would expect, the chief sea links are forged by those countries that have most of the following characteristics:

1   strong interest in overseas trade;
2   strong interest in overseas empire-building;
3   strong willingness to provide a shipping service for other nations;
4   strong interest in linking different coastal areas together within the same country;
5   harbours that can be used throughout the year (which means that harbours that freeze in winter are at a disadvantage);
6   harbours that can accommodate large modern ships (these are the so-called *deepwater* ports).

Figure 3-4 illustrates the location of the main shipping lanes in Eurasia. You will notice that the main lanes are all located in western Europe, the Mediterranean, and southern and eastern Asia. The Baltic areas, together with the entire Arctic coast and the Pacific north of Japan, are thinly served. Ice is the chief reason, although the U.S.S.R. tries to keep the lanes open for as long as possible by using the *Lenin* nuclear-powered ice-breaker. Another reason for the thin service throughout the northern half of Eurasia is the lack of trading possibilities; but this could be overcome if only the ice was not such a barrier. After all, the vast mineral wealth of Siberia could well benefit from the availability of cheap ocean transport. (In

The *Lenin* ice-breaker is one of only four nuclear-powered ships that have ever been built (not counting submarines). The others are the *Savannah* (American), the *Otto Hahn* (West German), and the *Mutsu* (Japanese).

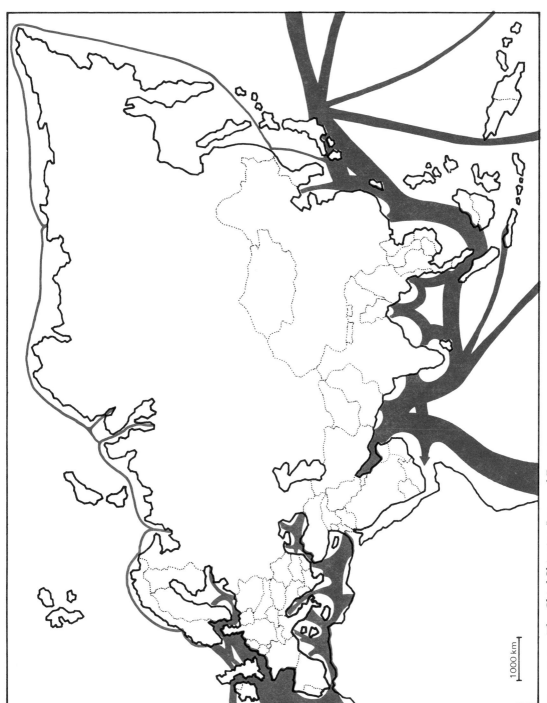

Figure 3-4.   The Chief Shipping Lanes of Eurasia

passing, do you remember the voyage of the *Manhattan* through Canada's northern waters in 1969?)

Looking more closely at Figure 3-4 you will see that, within the European area, the chief zone of shipping activity is the stretch of water from Denmark and Norway in the east to Ireland and Spain in the west. Basically the origin of most of this activity is the densely populated lands around the southern part of the North Sea, namely England, West Germany, the Netherlands, Belgium, and France. Most of these countries meet the majority of the specifications already listed. They all have had a strong interest in overseas trade and in overseas empire-building (especially Britain, which is the only *island* in the group). They all have harbours that can be used throughout the year, thanks to the Gulf Stream Drift. And they all have harbours that can accommodate the largest ships afloat. The chief harbours in this busy region are Rotterdam-Europoort, London, Hamburg, Antwerp, Bremen, Le Havre, Dunkirk, and Southampton. Rotterdam-Europoort is the busiest port in the world, handling most of the import trade of the European Community, especially oil, as well as a large part of the export trade. Figure 3-5 shows the exact locations of these major ports, as well as the locations of many of the second-rank ports. It also shows some of the internal ferry links within the region.

Because of the intense activity generated within this region, some other countries in Europe have developed a strong interest in attracting some of the carrying trade into their own ships. The most successful have been those countries which had already built up a maritime tradition by developing sea links between different parts of their own territory. Norway is the best example, closely followed by Greece. Both of these nations act as carriers for other countries, and of course they earn money by doing so.

The Mediterranean is a busy sea. It used to be even busier; it may even become busier again. But that depends on just when, and for how long, the Suez Canal is open for use. Whenever the Suez Canal is closed the Mediterranean is on its own; it is not a throughway. Nevertheless, even on its own it is busy, chiefly with oil tankers, liquid natural gas (LNG) carriers, and general freighters crossing from the north African coast. Libya, for example, sends mostly oil and dates; Algeria sends LNG and dates; Tunisia sends phosphates and dates.

In southern Asia the dominant trade is in oil from the

Rotterdam handles *at least* 250 000 000 t of imported freight each year. Toronto handles about 10 000 000 t *at most.*

The Suez Canal was opened for business on November 17, 1869. It is about 160 km long, with a maximum depth of about 10 m and a minimum width of some 60 m. Before it was closed by the Arab-Israeli war of 1967 it carried over 20 000 ships a year; it opened for use again in 1975, and it is now being deepened and widened to accommodate the larger ships now afloat.

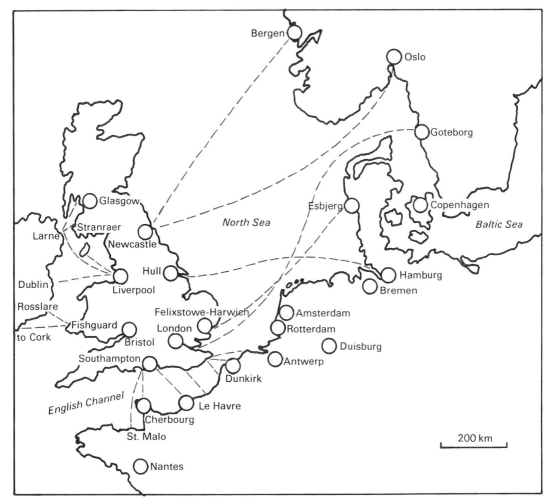

Figure 3-5. The North Sea-English Channel Region

Persian Gulf. The oil is shipped eastwards to India and Japan (by the way, can you now see why Kendo Bengoshi was selling ships to the Middle East?), and southwards round Africa to Europe and North America. Figure 3-6 gives you a simplified idea of the directions of the oil trade from the Middle East. The oil tankers are joined by lots of general freighters throughout southern Asia, because there is trade generated by India (chiefly through Bombay, but also to a lesser extent through Calcutta), by Malaysia (almost all through Singapore), and by Indonesia (through Djakarta). The products involved in all this trade are varied, including such items as cotton and tea from India, rubber and tin from Malaysia, and oil and rubber from Indonesia.

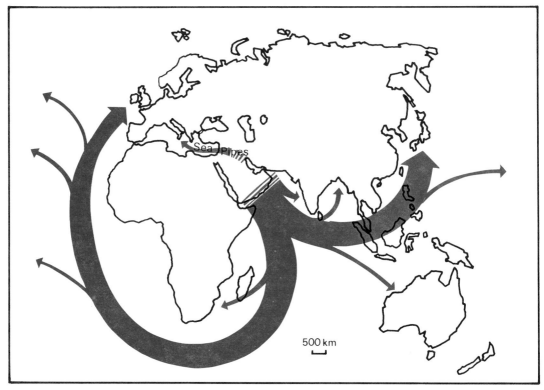

Figure 3-6. Simplified Oil Distribution from the Middle East

In the Orient, Japan dominates the trade scene. It meets *all* the specifications listed earlier, though its empire-building days are probably now over. A maritime tradition started early in Japan, because the country is made up of innumerable islands. In addition, people lived close to the sea (mentally as well as physically) because most of interior Japan was mountainous. Also, when Japan started on its deliberate economic growth in the 1870s it based much of its growth on foreign trade. It even built the ships for this trade; so successfully indeed that it now builds ships for half the world. And why stop there? Japan now has such an interest in ships and shipping that it advertises around the world for traffic: Please use our Japanese ships for carrying your products! In addition, of course, because of the warm Kuro Siwo current, Japanese harbours do not freeze. Nor are they limited only to small ships; the water is generally deep, because of the mountainous nature of the land. Japan has all the advantages.

There are over 3000 islands in Japan. Most people live on the four largest: Honshu, Kyushu, Shikoku, and Hokkaido. Over 80% of all the land is classed as mountainous. The highest and most famous mountain is Mount Fuji. Its height is 3776 m.

RAIL TRANSPORTATION

Railways were invented in Europe — in northern England to be exact. The first successful run was made in 1825 between Stockton and Darlington, a distance of about 20 km. Railway building soon developed into something of a mania, and by the 1880s there were railway lines spreading out over most of Europe. Gradually the uneconomic ones went bankrupt, and the more important ones were taken over — nationalized — by the various national governments. In fact, nowadays all the major rail systems in Europe are nationalized, such as British Rail in the U.K., Ferrovia Stata in Italy, and S.N.C.F. in France. The individual systems usually co-operate with one another in linking places within different countries, and there is now a very efficient arrangement of Trans-European Express routes. Trains are numerous and usually fast. Freight is similarly well organized, and freight cars are treated as part of an international pool, shared by all member countries.

Figure 3-7 shows you that railways are an available form of transportation in all countries, but it does not tell you much about the intensity of use they get. However, you can get some idea of the variations in use from Figure 3-8, which illustrates the quantity of tonne-kilometres. A tonne-kilometre is a unit that measures the amount of use a freight transportation system gets in a year. One tonne transported a distance of one kilometre is a tonne-kilometre; one tonne transported two kilometres is two tonne-kilometres; two tonnes transported one kilometre is also two tonne-kilometres. The use of a transportation system by passengers is measured in passenger-kilometres. (If 500 passengers are transported 150 km, how many passenger-kilometres of use are produced?)

Figure 3-8 shows 3 great concentrations of rail freight activity; Europe, including the U.S.S.R., is by far the largest, with smaller concentrations in India and the Orient. This is, of course, a strong reflection on the distribution of economic power; but it is not purely so. For example, Japan is a much stronger power than its use of railway freight indicates. The reason for Japan's low standing in railway freight use is that most of Japan's industries are on the coast, relying upon shipping rather than railways. Indeed, in Japan the railways are used much more for transporting people than freight; Japanese railways carry more people than the railways of any other

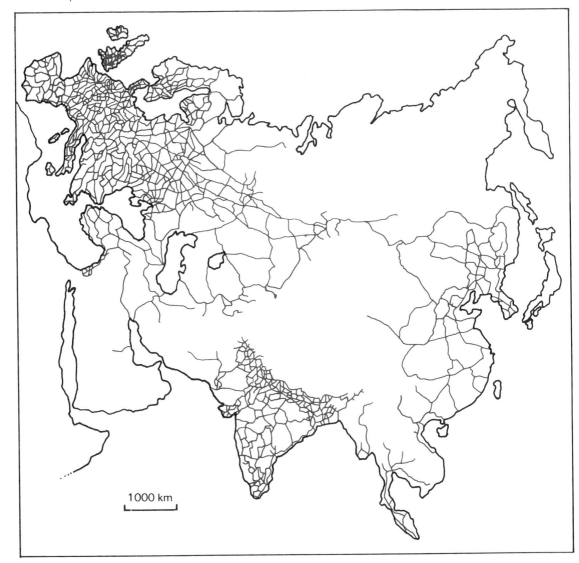

Figure 3-7.   Simplified Eurasian Rail Net

country. Japan's passenger-kilometre figure is about
300 000 million, compared with a little over 270 000 million for the U.S.S.R., the second greatest passenger-carry-
ing country. Little wonder, then, that it should be Japan
that pioneered special express passenger trains between
its major cities, particularly the "bullet train", or Tokaido
Express, that Kendo took.

One fact that stands out in Figure 3-7 is the essentially
localized nature of railway links. Links are made between
parts of Europe, between different parts of India, between
different parts of China, and so on, but links are not made

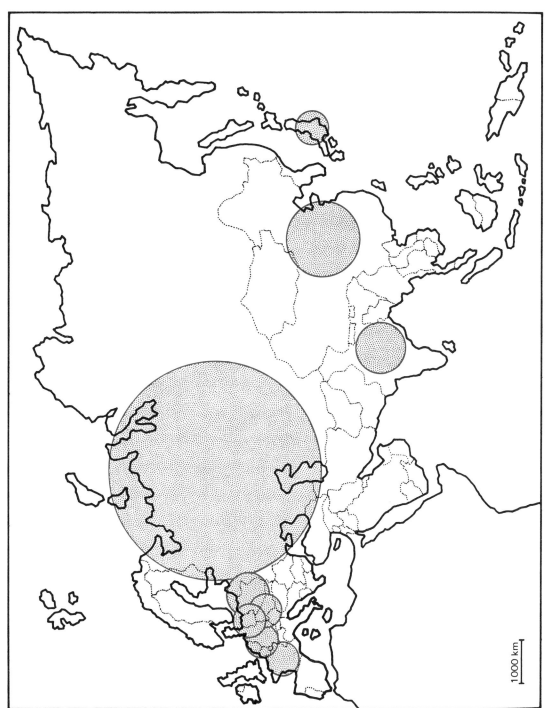

Figure 3-8.   Comparative Rail Freight: Chief Countries

1000 km

between the different major regions themselves. The big exception is clearly the Trans-Siberia Railway. Even so, not many people (nor much freight) actually travel along the 8-day route from Moscow to Vladivostock; mostly, the traffic is taken on and dropped off at various points along the way. Thus a person may travel from Chita to Irkutsk only, or from Omsk to Novosibirsk only.

ROAD TRANSPORTATION

Roads offer much more localized transportation than railways, but at the same time they offer much greater penetration within each region. You could drive pretty well anywhere within Europe by road, but hardly anyone would consider driving across Asia. The only people who like the idea of driving from, say, Europe to Singapore are testers for European and Japanese car manufacturers, who want to impress upon everyone the total reliability of their cars for everyday suburban driving.

An indication of the use of road transportation can be gained from Figure 3-9, which shows the ownership of cars and trucks in the different countries of Eurasia. Again, Europe stands out as an area with a high concentration of use (as reflected by ownership).

AIR TRANSPORTATION

The great concentration of activity in Europe stands out very clearly from Figures 3-10 and 3-11. The reasons are the ones you might by now expect: a strong interest in foreign trade; a strong interest in foreign travel; a set of strongly developed economies capable of supporting the cost of air transportation; and also, in the case of the U.S.S.R., a great need to bind together vast distances quickly and efficiently.

Because air transport is rather like ocean transport in one way — its ability to provide links over vast distances — it has also acquired another characteristic of ocean transport: the existence of what might be called *service carriers*. These are carriers which do not necessarily carry the passengers and freight from their own countries, but rather carry the passengers and freight of other countries. Just as Norway and Greece act as service carriers in ocean transport, so do the Netherlands, Belgium, and Switzerland in air transport. Can you name their national airlines?

Japan National Tourist Organization
Japan's railways carry more people than the railways of any other country.

Can you identify the following airlines and their country of origin?
JAL   TAP   KLM   PIA
BA   Sabena   Lufthansa
Aer Lingus   El Al   Aeroflot
MAS   SAS   Iberia AL
Alitalia   Finnair   KAC
PAL   SIA

Asia		Europe	
Afghanistan	50	Albania	n.a.
Bangladesh	n.a.	Austria	1 600
Burma	60	Belgium	2 400
China	n.a.	Bulgaria	n.a.
Hong Kong	130	Cyprus	70
India	1 150	Czechoslovakia	1 100
Indonesia	370	Denmark	1 300
Iran	350	Finland	900
Iraq	110	France	16 000
Israel	230	East Germany	1 500
Japan	18 000	West Germany	15 000
Jordan	20	Greece	400
Khmer	40	Hungary	400
Korea	130	Ireland	500
Kuwait	200	Italy	12 000
Laos	12	Luxembourg	110
Lebanon	140	Malta	60
Malaysia	370	Netherlands	3 000
Mongolia	n.a.	Norway	950
Nepal	10	Poland	800
Pakistan	150	Portugal	700
Philippines	450	Romania	80
Saudi Arabia	120	Spain	3 200
Singapore	180	Sweden	2 500
Sri Lanka	130	Switzerland	1 500
Syria	50	U.K.	14 000
Taiwan	100	U.S.S.R.	n.a.
Thailand	300	Yugoslavia	900
Turkey	300		
Vietnam	100		
Yemen	12		

Note: All figures in thousands (000)

Figure 3-9. Car and Truck Ownership in Eurasia

PIPELINE TRANSPORTATION

Pipelines are most useful for transporting goods that meet as many as possible of the following specifications:

1   Fluid handling. Natural gas is the easiest product to pipeline; oil thickens as temperatures fall, and oil pipelines need to be kept warm in cold areas. Water pipelines suffer from the same problem, and so, obviously, do slurry pipelines. (Slurry? What's that?)

2   Bulk production and demand. It is not much use to pipeline small quantities. For one thing a pipeline will not work at all unless it is *full*. Furthermore, the cost of construction is so high that it is only worth doing if you have large quantities of commodity throughput.

3   Steady production and demand. If production or demand are very uneven throughout the year, then it might be more economical to build great storage facil-

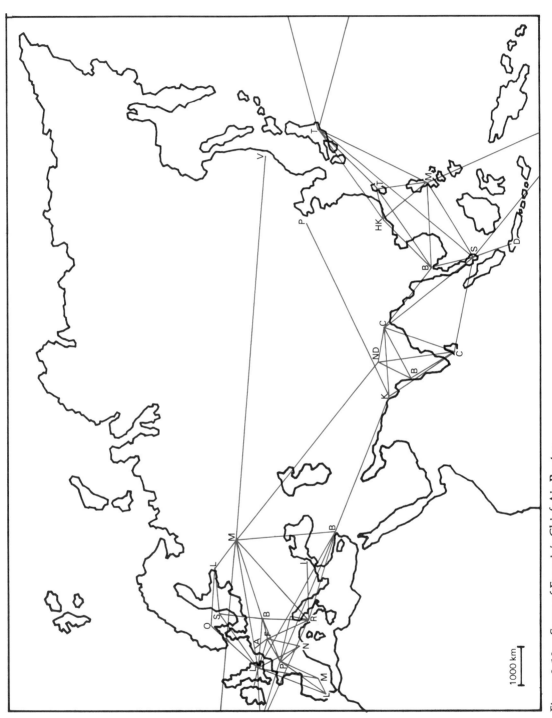

Figure 3-10. Some of Eurasia's Chief Air Routes

Asia		*Europe*	
Afghanistan	8 000	Albania	n.a.
Bangladesh	n.a.	Austria	7 000
Burma	2 300	Belgium	190 000
China	n.a.	Bulgaria	5 000
Hong Kong	n.a.	Cyprus	2 500
India	100 000	Czechoslovakia	13 000
Indonesia	17 000	Denmark	65 000
Iran	9 000	Finland	22 000
Iraq	2 000	France	500 000
Israel	110 000	East Germany	25 000
Japan	400 000	West Germany	500 000
Jordan	1 500	Greece	35 000
Khmer	500	Hungary	6 000
Korea	8 000	Ireland	60 000
Kuwait	15 000	Italy	300 000
Laos	800	Luxembourg	600
Lebanon	150 000	Malta	2 500
Malaysia	10 000	Netherlands	400 000
Mongolia	n.a.	Norway	65 000
Pakistan	70 000	Poland	7 000
Philippines	25 000	Portugal	40 000
Saudi Arabia	5 000	Romania	7 000
Singapore	10 000	Spain	100 000
Sri Lanka	2 000	Sweden	200 000
Syria	2 000	Switzerland	200 000
Taiwan	23 000	U.K.	550 000
Thailand	10 000	U.S.S.R.	2 000 000
Turkey	5 000	Yugoslavia	7 000
Vietnam	6 000		

Note: All figures in thousands (000)
Figure 3-11. Air Freight in Tonne-kilometres

ities and ship out by rail as required. But if production and demand are fairly even, pipelines can carry the commodities, continuously, straight to market, with only very small storage facilities.

Pipelines themselves offer a number of advantages in use. They don't require much of a "crew". They never crash; they don't go on strike; and there is no need to return the "empties" (cars, cartons, or whatever).

The map in Figure 3-12 shows the distribution of major pipelines in Eurasia. The concentration in Europe is again evident, with a lesser concentration in the oil-producing Middle East (where most of the oil is actually exported by ship). The Japanese, who require oil in bulk (and regularly), currently get most of it by tanker from the Middle East. They would like to get more of it by pipeline from western Siberia, but working out an arrangement with Russia is likely to take a long time.

The flow of a pipeline can be turned around simply by reversing the direction of pumping. In this manner the material in the pipe can be pumped either way according to the needs of the situation.

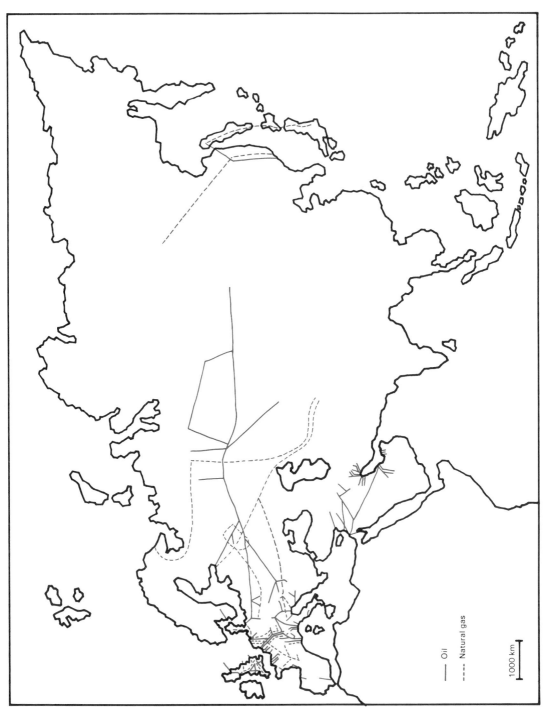

Figure 3-12. Important Pipelines in Eurasia

CONCLUSION

There are certainly 2 and possibly 3 levels of spatial interaction mentioned in this chapter. One is the major regional level, where the different transportation systems are best adapted to serving the needs of the different peoples *within* the major region. These transportation systems are certainly river, canal, and road, and possibly also rail and pipeline. However, both rail and pipeline could (and occasionally do) serve the interest of people in different major regions. For example, the Trans-Siberian Railway occasionally serves the needs of people to move from Europe into Asia; it certainly possesses the permanent potential to do so. So would a pipeline from western Siberia to Japan, although only for oil; but if an agreement is reached then a pipeline would certainly allow a major regional interaction.

Another level of spatial interaction is that *between* major regions, as supplied by ocean and air transportation systems.

QUESTIONS AND EXERCISES

1. There are 2 main types of international airport: traffic generators and staging posts. Can you name some examples of each type? What are the major factors involved in the location of both types of airport?
2. What are the relative advantages of air transport over ocean transport for the moving of people? And of ocean transport over air transport for the moving of freight? Does air have any advantage over ocean in the moving of freight?
3. What are the advantages and disadvantages of road transportation?
4. What factors limit the more widespread use of railways as transportation for people and freight?
5. Some people consider water a barrier to be crossed, and so they construct bridges and causeways and organize ferry services. Other people regard water as a transportation medium, and they look upon the bridges and causeways as obstacles. Can you find any examples, in Eurasia or elsewhere, of waterbodies that act both as barriers to be crossed by bridges and ferries and as routeways to be used by ships?
6. What does man have to do to improve the transportation aspects of naturally provided air, water, and land? Why does he have to do more with the land than with the air or water?

4. THE EUROPEAN COMMUNITY

CHOICE

Wars, wars, wars, and more wars. Europe's history is full of wars. World War I was largely fought there. World War II started there, and hundreds of other wars have been fought on the fields and in the towns of Europe, going clear back to the beginnings of history. Part of the problem has undoubtedly been caused by the small amounts of land available in Europe. There are large numbers of people to be supported in these relatively small spaces. Consequently, there has been a constant struggle between groups of people for control of what land there is. Another reason has probably been the fragmented nature of the European land mass. This may have encouraged the development of separate, and eventually conflicting, nations. For example, the Rhine River separated the Germans and the French, who continually fought each other for control over the waterway. The Alps separated France and Italy, and the French and the Italians fought for control of the Alpine passes. Over a period of many centuries, the national boundaries in Europe have shifted first one way, then the other, as states have fought and won (or lost). Some states have risen out of crumbling empires, as Norway did when it separated from Sweden in 1905. Still others have disappeared off the map altogether at various times, as Poland occasionally did.

An indication of the pressure exerted by different nations on the amount of land available is given by the figures for the average size of countries in each continent. In North America, the average size of the countries that occupy the continent is almost 10 000 000 km², but in Asia it is down to a little over 1 000 000 km². The continent with the smallest average size for its countries, however, is Europe, where the average size is under 300 000 km².

It is not only the large number of fairly small countries that sets the stage for war. These same countries are also among the most densely populated in the entire world. For example, the average population density for North America is 12 people per square kilometre, for Asia it is 75 people per square kilometre, and for Europe it is about 100 people per square kilometre. So from the point of

World War I, 1914-1918
World War II, 1939-1945
Other well-known wars in Europe's past include The Hundred Years' War, The Thirty Years' War, The Seven Years' War, the Napoleonic Wars, The War of the Austrian Succession, The Franco-Prussian War, The Austro-Prussian War, The Wars of the Roses, The Norman Conquest, The Swedish Wars, The Wars of Russian Expansion, and The Austro-Turkish Wars, just to name a few.

Chopin wrote much of his piano music under a strong *nationalistic* feeling. Some of his most impassioned works are the Polonaises, written to honour Poland when it no longer existed as a nation. Listen to the Polonaise in A Major, or the one in A flat, to get some idea of his strong feelings for his country.

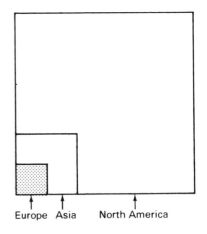

Average Size of Countries in Different Continents

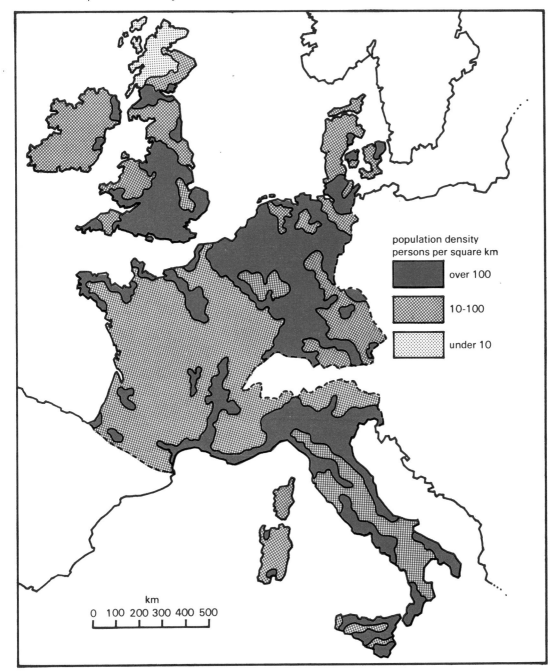

Figure 4-1.   Population Densities in the Common Market

view of having a high war risk Europe has the worst of
both worlds: it has the smallest average size of country
and the highest average population density. And so it has
had a lot of wars.

Even today, Europe is not without its trouble spots. There are still many small culture groups that do not want to belong to existing countries. They want independence, even though that means adding to the number of small countries. For example, the Bretons of northwest France want independence from France. The Scots and Welsh each want independence from the English. The Basques want independence from the Spaniards. The Northern Irish are fighting among themselves to decide whether they want independence from England. The Flemings and the Walloons fight each other inside Belgium, the Serbs and the Croatians want to break up Yugoslavia, and so on.

Nevertheless, things have changed. At least the big European countries do not fight each other now as they used to. The change came because a determined group of people wanted it to come, and they worked hard to achieve it. One of the hardest workers was Schuman, the French foreign minister who in 1950 put forward the Schuman Plan:

The uniting of the European nations requires that the age-old opposition between France and Germany be ended: the action to be taken must first of all concern France and Germany. . . . The pooling of coal and steel production will immediately assure the establishment of a common base for economic development, which is the first state for a European federation. . . . The community of production which will in this manner be created will clearly show that any war between France and Germany becomes not only unthinkable but in actual fact impossible. . . . By pooling basic production and by creating a new high authority whose decisions will be binding on France, Germany, and the other nations who may subsequently join, this proposal will create the first concrete foundation for a European federation which is so indispensable for the preservation of peace.

Schuman was not alone, however. Nor was he the first. But he was an influential voice at a time when many people throughout Europe were sickened by war. The first practical attempts at economic co-operation were in fact planned by the governments of BElgium, the NEtherlands, and LUxembourg during World War II. The plans were drawn up for a BENELUX union and signed at the Treaty of London in 1944. When the war ended in 1945, the

Marilyn Olsen

Crowded Europe: the rooftops of Florence.

The Basques?

The Flemings? Walloons?

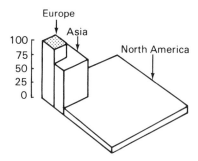

Average Population Densities
(people per square km)

In other words, if they share the making of steel then one of them is not likely to be making guns and tanks without the other one knowing about it.

governments continued to plan for a union of the 3 countries, and in 1948 the BENELUX union was actually established.

Meanwhile, in 1947 the U.S. government had started to supply gifts of money and equipment to Europe to help Europe repair the damage done during World War II. This was called Marshall Aid, and one of its conditions was that the countries of Europe should co-operate with one another in their recovery programs. The administration of the Marshall Aid recovery program was supervised by the Organization for European Economic Co-operation (OEEC), which was one of the world's more successful international organizations. The OEEC has now developed into the Organization for Economic Co-operation and Development (OECD), and it is still going strong.

Further attempts to get the countries of Europe to co-operate with one another instead of fighting were made in 1949, when the Council of Europe was established. This was set up as a sort of European parliament, where the representatives of the different countries could explain their views to one another. It was located at Strasbourg, which is on — of all places, considering its past history as a battleground — the River Rhine.

So you can see that in the late 1940s, after World War II, there were some strong movements towards co-operation and unity. Schuman harnessed them all together. He introduced a down-to-earth plan for avoiding future wars — a plan to combine the coal and steel industries of France and Germany. With this system, neither country could make steel weapons without the other's knowing about it, nor could either easily continue to do so during a war between them. As a result of Schuman's efforts, France and Germany were joined by Italy and the Benelux union in the signing of the Treaty of Paris in 1951. This established the European Coal and Steel Community (ECSC), which resulted in the pooling of all coal, iron, scrap, and steel production throughout the 6 members. It was as if there was just a single country producing coal and steel.

So great was the success of ECSC that the original workers and planners were jubilant. Doubters became supporters, and disbelievers moved up to become doubters ("Maybe there is something after all in this idea of union . . . ?"). Many people in Europe even came to regard themselves primarily as Europeans, and only secondarily as French or German or Dutch people. There was a grow-

OECD was reconstituted from OEEC in 1961. It now includes members from parts of the world other than Europe. Full members are Australia, Austria, Belgium, Canada, Denmark, Finland, France, West Germany, Greece, Iceland, Irish Republic, Italy, Japan, Luxembourg, the Netherlands, Norway, Portugal, Spain, Sweden, Switzerland, Turkey, U.K., and U.S.A.

ing sense of success, after years of war and economic disaster. Pressure began to mount for a wider and stronger union. Propaganda towards a wider union stressed the advantages:

1 Military conflicts would be ended forever.
2 European countries would regain their self-respect in the eyes of the world.
3 The rest of the world would pay more attention to the combined voice of the European nations.
4 Working conditions and living standards would be improved by joint action.
5 Increased production of goods would be encouraged by the disappearance of numerous trade barriers, thereby encouraging economies of scale.
6 More rapid technological progress would be brought about by co-operation.
7 Less-favoured parts of Europe could be helped.
8 Eventually a United States of Europe might be created.

Only a few people pointed out what they regarded as disadvantages — chiefly that union would cause a loss of national sovereignty. Most people ignored this, because they thought that too much attention to national sovereignty had been the prime cause of all the trouble in Europe over the centuries. So the movement towards union became a popular one, supported by the majority of the people. Not that everyone cared, though. Many people in Europe did not care at all; after all, politicians had talked before.

Nevertheless, in 1957 2 treaties were signed in Rome. The first Treaty of Rome set up the European Economic Community (EEC), which began in 1958. It quickly became known as the Common Market, and proved such a success that other countries, notably the U.K., soon wanted to join. Oddly enough, France was the chief opponent of other countries joining at that time. But then Schuman was no longer in authority. The second Treaty of Rome founded the European Atomic Energy Community (EAEC), which soon came to be called EURATOM. Before the end of the 1950s, therefore, Europe had gained 3 major overlapping unions: the ECSC, the EEC, and the EAEC, all with the same 6 members. (Which were . . . ?)

The consolidation of 6 different economies into one was not easy. There were many problems over tariff barriers, over rates of pay, over standards of production, over minority positions, over policies regarding ex-colonies, over qualifications, over laws — over millions of things.

A *mnemonic* is an aid to the memory. One kind of mnemonic is an *acronym*, made by combining the first letters (or groups of letters) of different words into just a single word, so that it is easier to say and to remember. You can then fairly easily break the made-up word (*acronym*) into the different items you want to remember. For example, you could use the *acronym* BENELUX to remember that Belgium, the Netherlands, and Luxembourg were joined in the first post-war customs union. Or even IKSA-QUAEO to remember the Arab States of the Persian Gulf. Try to make up one for the members of the European Community.

But the will was there to make co-operation work. People were forging a new Europe; they had high ideals, and much hope. In 1967 they took the next step. They combined the 3 existing communities (ECSC, EEC, and EAEC) into a single community: the European Community (EC). After a few more years of consolidation, and after de Gaulle had died, the European Community felt it could admit some new members. Thus in 1973 the Community expanded from 6 to 9 by taking in the U.K., Ireland, and Denmark, which had all tried unsuccessfully to get in on two previous occasions (1963 and 1968). Strangely, the enlarged EC soon began to run into trouble. Changes in governments often put national sovereignty ahead of co-operation; the Arab oil boycott of the Netherlands in 1973 caused severe strains; some of the community pioneers began to lose influence; and the U.K. wanted to renegotiate its terms of entry, because its new government thought that the U.K. had obtained a poor deal. In 1974 the German Chancellor even felt so down about the EC

1973
Britain, Ireland,
Denmark

Figure 4-2.   **Common Market Members**

that he described it as "a sorry tale of in-fighting and stagnation".

Despite the gloom that started in 1973-74, the European Community has accomplished a lot. Of course, it has not achieved everything it set out to, because full economic integration is an enormously difficult project. Besides, there are bound to be times when things do not seem to be going well. Nevertheless, if the will exists, and continues to exist, then the aims of the Paris and Rome Treaties will eventually be met. Let's examine what the EC is trying to achieve, and how far it has moved in that direction.

In general terms the aims of the EC are:

1   to promote the harmonious development of economic activities;
2   to promote a continuous and balanced economic expansion;
3   to promote a more stable economy;
4   to promote a faster rise in living standards;
5   to promote closer relations between the member states.

These are only general aims, however, and they need to be made more specific if any particular action is to be taken on them. For example, one of your general aims in life is undoubtedly to do well at geography in school. But in order to achieve that aim, you will have to do specific things, such as pay attention in class, learn your notes, pass your tests, hand in neat work, and so on. It is just the same with the aims of the EC; specific things have to be done if Europe is to succeed.

THE COMMON EXTERNAL TARIFF (CET)

One of the chief barriers to trade is tariffs, which list customs duties to be paid on goods brought into a country. Countries often set up tariff barriers against the goods of other countries in order to protect jobs or firms at home. By doing so, however, they deny their own people access to cheaper goods (or a wider selection of goods) from elsewhere. Tariff barriers, therefore, protect some people in a country at the expense of the majority, and they invariably reduce the amount of trade that takes place between countries. This is not good, because it stops countries from specializing in what they do best. When countries can specialize in what they do best, and trade freely with other countries that are also specializing in

Talking about aims and directions, try this: If you're standing at the North Pole and you then travel south down the 30°W longitude line to the 60°N latitude line, where you turn east for 60°, what direction will you then have to travel if you want to get back to the North Pole by the shortest possible route?

A tariff is a list of items with the costs or charges noted against each other.

A customs duty is a charge made on an imported good. For example, if you bought a camera outside Canada and tried to bring it back into Canada, you would have to pay a customs duty on it when you came through "customs".

Tariff barriers exist whenever a country makes its charges on customs duties so high that people are not prepared to pay them. They may try smuggling instead, of course. It happens!

what they do best, then goods can be produced more
cheaply, and everyone is better off. It is therefore good
to specialize and trade. At one time (before the EC) the
countries of Europe had tariff barriers against one another. People had to pay customs duties on goods that
crossed from France to Germany, or Italy to Belgium, or
whatever. The first thing that the Community set out to
do was to eliminate all these tariff barriers among its
members. This task was completed in 1968. As a result,
trade has expanded enormously within the EC. Figure 4-3
shows that trade within the Community (of Six, as it then
was) increased twice as fast between 1958 and 1970 as
did trade between the Community and the rest of the
world. (The internal EC trade index for 1970 of 575 is
more than twice the external EC trade index of 284.)
What is more, the actual value of EC trade, either internal
or external, *greatly* exceeds that of any other trading bloc
except the U.S.A. Even the U.S.A. now ranks third, after
Community trade with the rest of the world and Community trade within itself.

At the same time as the tariff barriers between the Community members were being eliminated, the various firms
in those countries (such as Fiat, Citroen, Philips, Siemens,
and Unilever) were able to treat the previously separate
markets of the different countries as just one market —
as a *common market*, in fact. However, the privilege of
being able to sell in a common market was restricted to
the members of the Community. Outsiders remained
exactly that — outsiders. All non-member countries still
had to face tariff barriers if they hoped to sell goods

	1958	1970	1970 Index (1958 = 100)
E.E.C. Internal trade	7 530	43 300	575
to world	15 910	45 200	284
to U.S.A.	1 660	6 630	399
to Canada	235	730	311
to non-member Europe	6 050	19 760	327
to Japan	140	990	707
Canada to world	5 050	16 180	320
U.S.A. to world	17 760	42 590	240
U.K. to world	9 110	19 350	212
Japan to world	2 880	19 320	671

Note: All values in $ millions

Figure 4-3. Data for Trade Growth

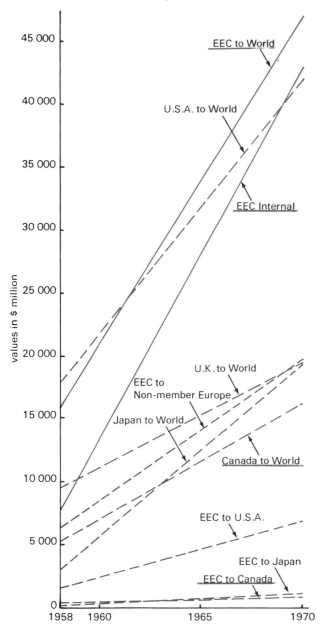

Figure 4-4.   Trade Growth, 1958-70

to the Community members. For example, customs duties had to be paid on Canadian wheat to get it into the market. Before the EC came into existence the countries that later became members all had different rates of customs duty, so that Canadian wheat, for example, could enter some countries more easily than others. Clearly, when the

Community internal tariffs were removed it made no sense to have a tariff wall around all the members that was lower in some parts than in others.

So what the members did, of course, was *harmonize* (or bring into common agreement) the external tariff. All members then had the same protection against outside competition, and outside goods could not be brought in more cheaply across one border than across another. Figure 4-5 illustrates the process of getting a Common External Tariff.

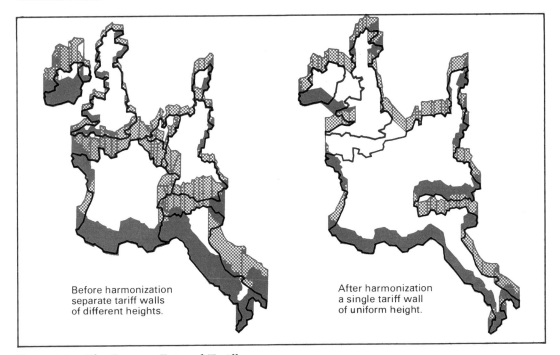

Before harmonization separate tariff walls of different heights.

After harmonization a single tariff wall of uniform height.

Figure 4-5.   The Common External Tariff

FREE MOVEMENT OF LABOUR AND CAPITAL

Economic opportunity is not — and never has been — equal in all parts of any one country, let alone throughout the entire Community. Certain areas tend to have natural advantages that favour economic growth, while other areas do not. You can probably readily think of some of these advantages. Access to transportation routes and sources of energy are two important examples.

If the opportunities presented in certain areas are to be fully developed, however, people must be free to move into those areas. This is called *labour mobility*. One of the aims of the EEC was to free the mobility of labour. Until this happened, countries faced two sorts of labour problem. First, they had to rely pretty well on their own populations to supply enough labour; if their economies were growing fast, this was a real problem. Second, they had to rely on their own opportunities to provide enough jobs for their populations, and this was also a problem if there were not enough opportunities. The idea behind freeing the mobility of labour was to make it possible for people who could not get jobs in one country to move to another country where there were labour shortages. In order to achieve this, the countries had to arrange that workers received the same social security benefits in the different countries, that wages were more or less similar, that medical care was equally available, that the jobs were advertised throughout the Community, and so on.

It was not easy; for some jobs, indeed, it has proved to be almost impossible. For example, lawyers find it extraordinarily difficult to transfer from one country to another, because the laws are often so different. There were all sorts of problems in other jobs too: different qualifications required, seniority rights, transfer of pension rights, shortages of housing, leaving of friends and relatives, and so on, not to mention the language problem. Nevertheless, since 1964 all Community workers have had equal status in being considered for jobs, and since 1966 they have all had equal social security rights with the citizens of the country they have moved to. As a result of these measures to make labour more mobile, there has been a massive movement of people from the outer fringes of the Common Market, especially southern Italy, into the areas of greater economic opportunity, especially West Germany, which is still very short of labour.

Capital movements — the flows of money — have always been fairly easy between nations, even though there have been times when certain countries have taken measures to control the flow of money. Not many countries object to foreign money coming in (although West Germany has done so at various times because of the danger of inflation), but a lot object to money going out of the country. Both the U.K. and Italy have frequently resorted to controls to stop money from leaving the country. But they both welcome money coming in. In the U.K. the banks of London can usually make a profit out of investing it. Italy welcomes foreign money because it can be used to help

Daphne Ross

An area of low economic opportunity in S. Italy.

People from southern Italy did not move just to West Germany. Many stayed within Italy itself, simply moving to the north of the country instead of staying in the south. Many others left Europe altogether, emigrating to Canada or the United States (or Argentina and Brazil). In Canada, most Italian immigrants settled in Ontario, especially in Toronto. In 1974 the Premier of Ontario made a special visit to Italy in honour of all the Italian immigrants in Ontario.

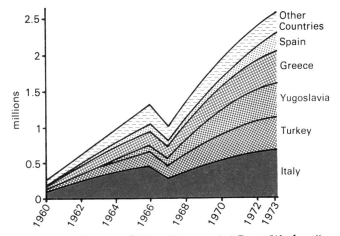

Figure 4-6.   Sources of West Germany's "Guest Workers"

develop the relatively poor south. Some of the foreign money moving around Europe from bank account to bank account is European money; in other words it is German *deutschemarks*, French *francs*, Dutch *guilder*, British *pounds*, Italian *lire*, Danish *kroner*, Irish *pounds*, or Belgian *francs*. Some of it is also Arab oil money, some Japanese or Hong Kong money, but most of it is U.S. dollars. The U.S. does not want these dollars back, because they would just make inflation worse at home; nor will it exchange them for gold (as it used to at one time). So the dollars are more or less stuck in Europe (they are now called *Eurodollars!*), where they move from country to country in search of the best economic opportunities. What would you do if you owned some of these Eurodollars? What some owners have done is buy up a few American companies in the United States; for example, Standard Oil of Ohio is now owned by British Petroleum. But that sort of thing does not happen very often.

THE COMMON AGRICULTURAL POLICY (CAP)

The Common Agricultural Policy is basically still a question mark. The idea behind it was that trade in agricultural products could never be achieved as easily as could trade in manufactured products. For example, cars can be made in Italy just as easily as in Germany, but dairying cannot so easily be equally practised. Agriculture is, of course, much more dependent upon suitable climatic factors than manufacturing could ever be. Not only that, but peasant-farming in most of Europe was a centuries-old way of life, and the organizers of the EEC were not so heartless as to want to destroy such a life-style for millions of people just by the signing of a Treaty.

If you put these two facts together you can see that the EEC organizers wanted to protect farmers who would become uncompetitive if free trade in agricultural produce was suddenly permitted. For example, since dairying is much easier (and therefore more profitable) in Germany than it is in Italy, the opening up of the Italian dairy market to German competition would have put many Italian dairy farmers right out of business. The EEC Treaty signers were not keen on that idea, so they decided on an agricultural policy that protected local farmers anywhere within the Community from cheaper competition from elsewhere, either from elsewhere within the Community or from anywhere outside the Community. Thus Italian dairy farmers are protected from cheaper competition from Germany, and German fruit growers are protected from cheaper competition from Italy, and all farmers within the EEC are protected from cheaper competition from, say, Canadian wheat farmers or New Zealand butter producers.

But the policy covered more than just that. The level of protection was fixed to support the relatively inefficient peasant farmers, especially those of France (because it was chiefly their life-style that the organizers wanted to protect). This in turn made it very profitable indeed for the *efficient* farmers to grow and sell much more food than ever before. After all, if the subsidies from the EEC common agricultural fund could support even an efficient farmer, think how much extra income an efficient farmer could earn by increasing his output!

The CAP has come under a lot of attack within the Community, particularly from those members which are not important farming countries. It has also come under a lot of attack from some important farming countries *outside* the Community. Can you see why the CAP should come under attack?

Let's see what the reasons are. First the causes of attacks from within the EEC:

1   The price of food is high, because the prices have to cover the cost of keeping the small and inefficient farmers in production.

2   The more industrialized countries, such as Germany and the U.K., resent having to pay out millions of

A centuries-old way of life:

Marilyn Olsen

·French farm near St. Etienne.

Marilyn Olsen

Ox-cart in the Rhone valley.

Marilyn Olsen

Olives and fields near Rome.

Marilyn Olsen

Danish farm-house.

dollars each year in order to keep inefficient farmers in production in other EEC countries.

From outside the Community, countries object because:

3   The subsidies offered to the already efficient farmers within the EEC are unnecessary. These farmers then produce more than they normally would, and this cuts down the export chances of other countries. For example, Canada cannot so easily still sell wheat to the EEC, not even to the U.K.

4   The huge surpluses that are created within the EEC, especially in those items where their efficient farmers are really *very* efficient (such as dairying), causes embarrassment to the EEC organizers, who often do not know what to do with all the food that is produced. Sometimes they feel they should give it away to the hungry people of Africa and India, but the food isn't always suitable for long-distance transportation and distribution in the Tropics. Sometimes they feel they should destroy it, which they occasionally do because they cannot easily store it. Sometimes they try to store it, as they have often done with butter and meat. And sometimes they sell it off cheaply on the world markets, which annoys regular world suppliers such as Canada, Australia, and the U.S.A. immensely, because it causes world prices to drop, and so Canadian farmers, etc., do not earn as much as usual.

So you can see that there are quite a few reasons for the CAP to come under attack. On the other hand, there are certainly no shortages of food within the EC, even though the prices are high. What do you think about it all?

Marilyn Olsen

A contrast in pastureland between the N. European Plain and S. Italy.

Daphne Ross

THE EUROPEAN ATOMIC ENERGY COMMUNITY (EURATOM)

In order to help achieve one of the main aims of the Community — the raising of living standards — a great deal of emphasis has been placed on the development of electric power. Generally, within the EC electricity production is doubling every 10 years or so, and people expect nuclear electricity to play an ever-increasing role in the total power play. By 1980, in fact, the EC expects to have about 25% of its electricity generated by nuclear means (from merely 1% in 1965). To this end, Euratom runs research centres at Karlsruhe in Germany, Petten in the Netherlands, Geel in Belgium, and Ispra in Italy (see Fig-

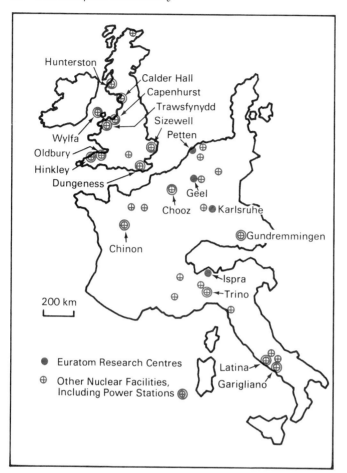

Figure 4-7. Nuclear Energy

ure 4-7). It also co-operates with individual members' own research centres. In addition to research it also oversees the generating programs that member countries have set up. The chief problem with all of this is that there is no agreed *European* program — just a lot of ill-combined *national* programs (as you could guess from the sort of secrecy that many people associate with nuclear power). In consequence, research efforts tend to be split, and a lot of money is wasted.

In one country alone — Britain — the nuclear research effort was so split that at one time there were hardly two reactors alike, and the nuclear power program was years behind schedule. In 1974 the problem was so serious that Britain decided to buy the CANDU-type technology from Canada.

EC RELATIONS WITH ASSOCIATED COUNTRIES

Under the terms of the Rome Treaty, association with the Community is open to *all* countries. However, not all

countries have sought to develop an association. Those
that have sought such an association have usually managed to achieve some sort of customs union, whereby
tariff barriers are removed between the EC and its associates. Sometimes, too, the EC gives aid to the associates
in order to help them develop their economies to such a
stage where they might consider applying for full membership. The greatest number of associated nations is in
Africa (see Figure 4-8). There, 18 of the former colonies
of France, Belgium, and Germany signed the Yaoundé
Convention in 1963, agreeing to the eventual formation
of a free-trade area and the interim receipt of aid. Nigeria,
though not a party to the convention, has also sought
special association. Individual trade agreements have also
been signed between the EC and Sweden, Switzerland,
Austria, Portugal, Norway, Iran, Lebanon, and Israel. The
closest associations, however, have been negotiated with
Greece and Turkey, and it is expected that both these
countries will eventually apply for full Community membership.

HAS THE COMMUNITY BEEN A SUCCESS?

Many things still remain to be done — developing an
acceptable CAP, obtaining regional equality, achieving a
single European currency, and so on. But there is no doubt
that the EC has been a huge success. At the very basic
level, meeting Schuman's main purpose, there has not
been a war between any of the major powers of Europe —
and that is quite important. At a different level there has
certainly been a great expansion in trade, as Figure 4-4
shows only too clearly. Living standards have certainly
risen, often at a very fast rate, although some members
have grown richer faster than others. The nations as a
whole have become richer, though: their industrial production is up, food supplies are up, and Gross National
Products are much greater. People have more cars, more
TVs, more holidays, and more food.

Marilyn Olsen

*People have more cars: traffic at
the Arc de Triomphe, Paris.*

   Of course, *all* industrialized nations have become richer
since 1958, not just the members of the European Community. Figure 4-9 shows index numbers for a variety of
countries, so that you can see just how well the EC members have done in comparison with some other important
countries. Index numbers, of course, only show comparisons in rates of growth. They do not show you how the

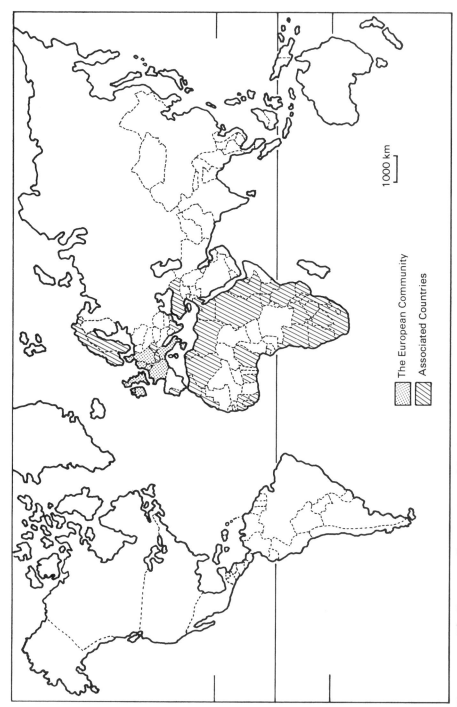

1000 km

The European Community

Associated Countries

Figure 4-8.   The European Community and Its Associates

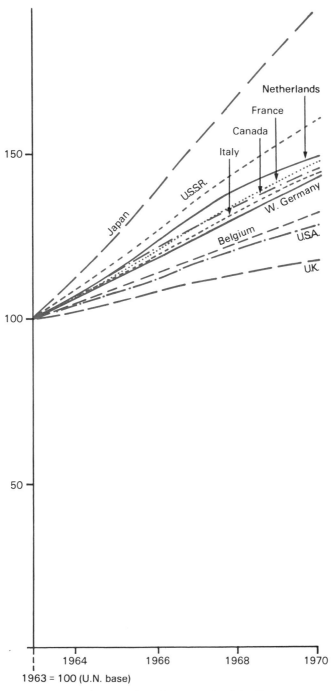

Figure 4-9. Index Numbers for Growth of GNP, 1960-70

	Population (millions)	Area (square km)	Total GDP ($ billions)	Per capita GDP ($)	Average annual % growth of GDP in real terms since 1961	Output of steel in millions of tonnes	Number of persons per car	Number of persons per physician	Number of persons per telephone
Belgium-Luxembourg	9	30 513	37	4111	4.8	13	4.3	644	4.0
France	50	547 026	196	3920	5.8	24	3.8	747	5.6
West Germany	60	247 973	257	4283	4.5	45	4.3	568	4.3
Italy	50	301 225	117	2340	4.8	18	4.5	553	5.0
Netherlands	12	40 844	46	3833	5.5	5	4.4	832	3.4
Denmark	5	43 069	21	4200	4.7	0.5	4.5	689	2.9
Ireland	3	70 283	5	1667	4.0	0	7.5	963	8.6
U.K.	56	244 013	154	2750	2.6	29	4.7	855	3.7
European Community	245	1 524 946	833	3400	4.5	134.5	4.4	670	4.4
Canada	22	9 976 139	110	5000	5.5	16	3.1	700	2.2
U.S.A.	205	9 363 353	1175	5731	4.4	130	2.3	669	1.7
Japan	105	369 881	295	2809	10.1	100	12	898	3.8
U.S.S.R.	245	22 402 200	n.a.	n.a.	7.5(e)	120	n.a.	433	22.0
China	800	9 561 000	n.a.	n.a.	10.0(e)	18	n.a.	n.a.	n.a.

n.a.   Not available
(e)   Estimate

Figure 4-10. Some Comparisons, using 1970-71 data

EC members actually stand in relation to other important countries. Figure 4-10 does, though.

QUESTIONS AND EXERCISES

1. Tariff barriers do have some useful functions; otherwise, countries would not have them. Can you think of (or find out) *three* good reasons for the existence of tariffs?
2. Why is it good for countries to trade more with one another?
3. What are the various non-trade advantages of the joining together of countries in the European Community?
4. Find out what you can about *Commonwealth Preference*. Canada used to trade with the U.K. on this basis. What do you think has happened to Canada-U.K. trade since 1973, when the U.K. joined the Community? Can you find out whether what you think is actually the truth?

5. EUROPEAN FARMING

AREAL DIFFERENTIATION

The landscape of the world's smallest continent is extra-ordinarily rich and varied. Snow-capped mountains alternate with rich lowlands, high plateaus with broad plains, polderlands with sandy heaths. All told, the land area of Europe, excluding the U.S.S.R., is some 4.9 million km². This may seem an enormous amount (roughly half of Canada's total land area), but it is only 3.3% of the world's total land area.

Surprisingly, nearly 500 million people live there. It is surprising because this figure is about 12% of the world's total population, making Europe about 3 times more densely peopled than the world average. Although closely followed by Asia, which is only a little less densely populated, Europe is the most densely peopled continent in the world.

Knowing that, you may think that Europe is crowded, perhaps unpleasantly so. Yet Europe has some of the world's highest living standards. There is plenty of food, many valuable raw materials such as coal and iron ore, numerous factories and highly developed services such as railways, roads, canals, seaports, and so on. Europe has such a high density of population precisely because it has the ability to support all these people.

When we talk about the support of life, we instinctively think of . . . Well? What are you thinking of? Food? If so, good. People are sustained very well by European food

It has been estimated that for a European farm to be profitable, a crop farmer needs 80 to 120 ha of land, a dairy farmer needs 40 to 60 cows, a beef producer needs to produce 150 to 200 carcasses annually, an egg farmer needs at least 10 000 birds, and a poultry producer needs to kill 100 000 birds a year.

Unit	Total land area	Arable land area	Pasture land area	Forest area	Other land area	Population in thousands
Europe	493 000	145 000	91 000	140 000	117 000	470 000
North America	2 242 000	271 000	362 000	824 000	785 000	320 000
South America	1 783 000	84 000	363 000	908 000	428 000	200 000
Asia	4 993 000	696 000	886 000	1 452 000	1 959 000	2 400 000
Africa	3 031 000	214 000	822 000	635 000	1 360 000	360 000
Oceania	851 000	47 000	463 000	82 000	259 000	20 000
Antarctica	1 510 000	—	—	—	1 510 000	—
Totals	14 903 000	1 457 000	2 987 000	4 041 000	6 418 000	3 770 000

Thousands of hectares

Figure 5-1.   World Land Use, 1971

production, even though on the face of it the land surface appears to be crowded. There are all kinds of reasons why this should be so, but the most important one is that Europe has an ample supply of farmland. Not excessive, nor even particularly abundant, but more than enough. There is enough so that, if it is used wisely and well, it can provide a great deal.

QUESTIONS AND EXERCISES

1. On one sheet of paper, construct a series of equal-area bars. There should be one bar for each of the units mentioned in Figure 5-1. If each bar is about 10 or 15 cm long and perhaps 2 or 3 cm wide, they should all fit on one sheet. Make sure these bars are aligned correctly, one above the other.

    Calculate the land-use percentage in each category for each of the units listed in Figure 5-1. Divide each bar accordingly.

    Write a paragraph or so comparing Europe's land-use pattern with those of the other units. In particular, what seems to have happened to Europe's forests?

2. On one sheet of graph paper, show as a series of proportional shapes the arable (cropland) areas available per person in each of the units listed in Figure 5-1. This time, include a shape for the world total, too. Now for the hard part: Give the dimensions of each proportional shape in metres. One ha covers an area of 10 000 m².

    Why is arable land so important?

3. Calculate overall population densities — that is, the number of people to each square kilometre of total land area in each of the named units in Figure 5-1. Don't forget a figure for the world total. One km² covers an area of 100 ha.

From the above assignments and questions, many things should have emerged. You should take special note that, although Europe is very densely peopled considering its *total* land area (arable included), in terms of *arable land alone* Europeans are just about at the world average density. Their continent is very fortunate in its climate types and the length of the growing season (about 8 months long in most areas). In addition the soils are fertile or can be made so by drainage, irrigation, or the application of fertilizers. Even so, Europeans have only their fair share of arable land, no more. Thus, in order to try to feed everyone, European farmers really have to work hard. As we have just indicated, nature helps, but the farmers have to make the best use of what nature offers.

Precipitation totals everywhere are at least 500 mm annually. However, in the Mediterranean lands the summer

Marilyn Olsen

Oranges growing in a Spanish grove.

drought creates problems and irrigation is often required for crops such as rice.

Cereals, such as rice, wheat, oats, barley, and rye, are common; *roots,* such as carrots, potatoes, turnips, and beets, are widespread; and *fruits,* such as apples, pears, plums, oranges, grapes, and lemons, are very important. A complete list of the types of crop produced in Europe would be almost endless, so we shall not attempt one here. However, bear in mind that most European farming is *cash farming.* This means that the crops are grown to be sold for money, which the farmer will then use in the marketplace just like anybody else. *Subsistence farming* — a more peasant-type of agriculture in which most of the food produced is eaten by the farmer and his family — is dying out but still survives in some of the more remote areas. By now most farmers are aware that selling their crops in the cities is the only way to make farming pay enough to provide money to buy gas for the tractor, paint for the house, and a holiday once in a while. A subsistence farmer could not hope to raise enough cash for such luxuries (or are they really necessities?), and the age of barter is almost dead. Therefore, the standard of living of a subsistence farmer is quite low by European standards, and that is why there are so few left. Most of them are in the Mediterranean lands, but only in the more isolated regions such as Peloponnesus in Greece, parts of Sicily, Sardinia, Corsica and parts of mainland Italy, remote areas of Spain and Portugal, and some of the outer islands of the British Isles.

Cash farmers utilize the best soils of Europe. Most of these soils benefited from the last Ice Age; a great deal of finely ground rock from Scandinavia was carried south by melting ice and dumped on what is now called the North European Plain. You can pick this out in the atlas as a broad belt of lowland that sweeps right across the northern part of Europe from northern France to the immensity of the U.S.S.R. To the south of this great plain, the land is more broken up by hill and mountain ranges, but there are still great lowland areas or *basins* that act as natural traps for moisture and fine soil particles washed down from the nearby hills. Such *alluvium* lies in an enormously thick deposit on the floor of the flood plain of the Po river in northern Italy. This is a fine example of a well-cultivated basin.

Working with the natural advantages that Europe has to offer, the farmer exploits them to the best advantage.

Have you ever heard of of kohlrabi? mangold wurzels? lucerne? They are all European *fodder* crops! (What is fodder?)

Marilyn Olsen

Olives grow on a xerophytic *type of tree. Note how narrow these leaves are. What else helps the olive tree conserve moisture?*

Bonnie Eccles

A Macedonian shepherd, his flock, and his faithful dog, moving at a pace that has not changed in thousands of years.

The Common Agricultural Policy of the European Community has resulted in the overproduction of several things. This is because much European farming is still done on smallholdings, which are not very efficient, but because there are a lot of smallholders, the governments of places like France and West Germany do not want to lose their votes. That would probably happen if the CAP was dropped and artificially high, guaranteed prices went with it. Then a lot of marginal farmers would go out of business. But because the CAP is in operation, they are guaranteed a living. Of course this has resulted in expensive surpluses being created. *Butter* has to be "dumped" in markets overseas and sold to old-age pensioners and students at half-price. Surplus *beef* is bought at the rate of 5000 tonnes a week by the governments of the EC countries. By the middle of 1974, a mountain of beef weighing over 150 000 tonnes had been overproduced, bought by the governments and stored at a daily cost of over $300 000.

Shell Photographic Service

European livestock is intensively raised. Animals are penned and fed a controlled, scientific diet. What advantages does this have over allowing pigs (hogs) to root about in a more traditional fashion?

In this, the farmers of Europe are fully backed by the resources of countries with high standards of living and *technology*. Through technology have been developed tractors and farm machinery, fertilizer, top-quality seeds and carefully bred animals, sophisticated transportation routes and equipment, agricultural research at colleges, universities and government laboratories, as well as a great many other things. Also aiding the farmer are government support policies for agriculture. These take the form of guaranteed prices under such things as the Common Agricultural Policy of the EC, and marketing boards for products such as meat, fish, eggs, wheat, potatoes, and milk. The purpose of marketing boards is to take the "boom-bust" cycle out of farming. In a good year, too many crops on the market would mean such low prices that many farmers would be forced out of business or, at best, not guaranteed a fair price for a hard year's work. On the opposite side of the coin, there might not be enough food to go around in a poor harvest year, so the consumer would suffer. Governments now buy up surpluses in times of plenty and dispose of them where they can, or they store them against the time when a poor year on the farms requires that the surpluses be used.

On the whole, farmers in Europe are given all kinds of encouragement to produce food and industrial crops.

Sometimes they are even paid by the government *not* to produce particular crops or even any crops at all on some parts of their farms, so that those fields can lie *fallow*, at rest, for a year or two. Fields left fallow will regain some of their natural fertility and moisture. Just how well do European farmers perform then, given all their special advantages?

Crop	World	South America	Central and North America	Africa	Asia	Oceania	Europe
Wheat	3116 / 210.3	82 / 6.8	487 / 23.8	77 / 8.7	478 / 43.4	83 / 7.1	671 / 27.3
Rye	308 / 20.0	4 / 0.6	15 / 1.0	no data	7 / 0.6	no data	131 / 6.7
Barley	1285 / 78.1	11 / 1.1	182 / 8.2	43 / 5.2	116 / 10.6	24 / 2.2	442 / 16.3
Oats	526 / 32.0	8 / 0.7	189 / 10.5	2 / 0.4	6 / 0.5	16 / 1.8	171 / 7.3
Rice	3068 / 135.5	105 / 5.6	54 / 1.5	75 / 3.9	1801 / 90.6	3 / 0.06	19 / 0.4
Potatoes	2995 / 22.5	90 / 1.1	179 / 0.8	28 / 0.4	132 / 1.3	9 / 0.1	1260 / 7.3
Cotton	118 / 33.2	11 / 3.8	28 / 5.3	13 / 4.6	25 / 11.4	no data	2 / 0.3

Note: Crop production in hundreds of
      millions of kilograms
                              crop area in millions
                              of hectares

Figure 5-2.   World Production Statistics for Some Internationally Important Crops, 1970

4. Which continent has the highest crop yields in the table given as Figure 5-2? To give impact to your answer(s), put them in the form of a graph. You may use any type of graph that you wish. If you do not have many ideas on this, ask your teacher to discuss some suggestions with you.

The basic difference between farming in Europe and the rest of the world is that once Europe's farmers decide to produce a particular crop, they usually do it very well when compared with the rest of the world. Why? Well, apart from modern technology and other aids that we have already mentioned, *competition for use of the land* is most important.

Wherever different uses compete for land, the price of land is forced upward. For example, at the edge of large cities, in what is known as *suburbia*, land is continually changing in use. Farmers who have often done nothing but work the land all their lives are approached with an offer. Will they sell? Usually the price offered is a good one, more than the farmer could have received if he had simply sold the land to another farmer. The high price offered reflects the intention of a developer to use the land for houses, shopping centres, and industrial parks. A crop of houses will bring much more money than a crop of wheat. As the use of the land has changed, so has its value.

Well, fine, you might think. No harm has been done. The original farmer is happy with his profit, and builders are happy because they can start building. Prospective house buyers are happy, too, because there are more houses to choose from. Just where is the weakness in the argument that city growth is good?

Let's open the atlas to a map of Europe. On what sort of land do most of the major cities appear to be built? Lowland! But as you know, it is on the lowland areas of Europe that the great majority of the continent's farming is done. The great cities of Europe simply had to be built in the centre of fine agricultural land. Otherwise, all those hundreds of years ago the townspeople could not have been fed.

For centuries, the cities of Europe remained fairly static in terms of size. Then, about 100 years ago, they mushroomed with the explosion in industrial activity that has come to be called the *industrial revolution.* Peasants flocked to the cities in search of work, education, and opportunities unavailable to them elsewhere. As the industrial revolution helped people to live longer, the population of Europe grew fast — and as it did so, most of those extra people found their way to the cities, if they hadn't already been born there. Today, three-quarters of Europe's people live in cities.

Naturally, as the cities grew, the supply of farmland diminished, and so did the number of farmers. Fewer people had to produce more food than ever *and* ship it to the cities. Of course, the industrial revolution helped the farmer as much as it did anyone else. Better seed, revolutionary new machinery, and improved transportation all helped. So did the increase in the price of food, production of which naturally lagged quite a lot behind the rate of increase of the population. There was a lot of incentive for farmers to produce more and more food, just as the supply of farmland was diminishing and the number of

Royal Danish Embassy

Poultry breeding during the past 40 years has become an important part of Danish agriculture. The production of eggs and dressed poultry has increased and Denmark now exports to a number of countries.

Region	1961	1962	1963	1964	1965	1966	1967	1968	1969	1970	1971	1972
World	98	100	100	102	101	103	106	107	105	106	108	105
Western Europe	96	101	101	102	101	103	109	111	109	111	114	113
Eastern Europe incl. U.S.S.R.	99	100	97	103	103	109	110	111	109	113	114	112
North America	96	98	102	101	103	105	109	108	107	104	113	109
South America	100	99	100	100	101	99	102	99	102	102	100	97
Oceania	94	102	102	105	97	110	99	117	110	106	111	107
Near East	97	102	100	101	100	102	103	104	104	103	101	106
Far East excl. China	101	100	101	102	96	95	97	100	102	105	102	97
Africa	97	100	102	101	101	97	99	100	102	101	102	101

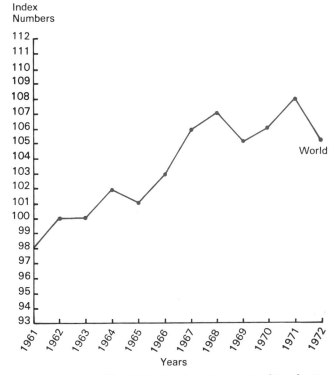

Figure 5-3.   World and Regional per Capita Food Production: Index Numbers (1961 – 1965 = 100)

mouths to be fed was multiplying. Farm sizes increased as peasants sold out and left. Holdings were therefore *rationalized*. This meant that individual farms took on larger and more regular shapes so that the land was easier to work efficiently with machinery. Above all, the cities provided a market for the food produced.

It is quite clear that today, if a farmer wants to start up in business, he must pay a lot for the land. He can only afford it if he has the skill and knowledge to make the farm pay well. How else could he meet his monthly mortgage payments? This means that the successful European farmer must use every available square metre he has. Once he has done this, then the only way he can produce more is to use modern technology.

5. Examine Figure 5-3 closely. It shows the rate of change of food production per person everywhere in the world. Here is how to read the table: If in 1960 every person in the world had 100 loaves of bread to eat, then in 1965, even though the world population had increased by perhaps 300 million people, food production had increased enough to provide 101 loaves of bread per person *if all the food were to be shared out equally*. This is what the words "per capita" mean — per person. But as we know all too well, food production is *not* shared out equally.

    If in 1965 the index number for per-capita food production had been 100 and not 101, then you would know that world food production had just kept pace with the increase in world population. And if the index number had fallen to 90, then you would know that world population had grown faster than world food supplies.

    Copy down your own graph as shown in Figure 5-3, and complete it for all the named areas in the world. How does Europe compare with the rest of the world? Why is it so consistent compared with places like Asia or South America?

6. European farmers are *intensive* farmers. What does this mean? How can intensive farmers raise production?

Above all, European farming is very *efficient* when compared with the rest of the world. But what about comparisons *within* Europe? Who are the most efficient farmers on the continent? And what do they produce most efficiently?

Farm organization plays a big part in the answers to these questions. In the communist countries of eastern Europe, very little land is privately owned. Great state farms have replaced large estates and the numerous peasant smallholdings. Since the collective and state farms have only been in operation in the eastern European coun-

By the end of the century, Poland hopes to increase agricultural production by 10 times.

tries for perhaps 25 years, there are still problems to be overcome in organization, marketing, and planning. We shall return to them later. Agriculture is only one part of any country's economy, and communist governments generally try to plan the entire economy. This is an immense task.

By way of contrast, a country such as the Netherlands does not have nearly so much government control in agriculture. Generally, the government control takes the form of making new agricultural land. The Netherlands is one of the world's most densely peopled countries. Each person has less than 0.25 ha of total land area. By way of comparison, Canada has 42.0 ha per person. Since much of the Netherlands is swampy and below sea level yet potentially very fertile, great banks of sand and clay — the famous *dykes* — are used to surround large areas, which are then pumped dry so that farming can commence. Land so reclaimed from the sea by *dykes* is known as a *polder*. Some polders are very large, but individual farms are very small, often less than 5 ha in size.

The Netherlands, as you know, is part of a much larger economic unit, the European Community. The 9 nations in the EC have a common agricultural policy, so that their farmers are protected from foreign competition. Everything they grow or produce has to be bought by their governments at artificially high prices. This leads to overproduction of things such as butter, because farmers find it so profitable to produce. Consequently, many marginal farmers who would otherwise go out of business still continue. However, EC countries still maintain a high standard of agriculture. It is just not quite as efficient as it might be, that's all.

The Mediterranean countries form quite a distinct farming group. Countries such as Italy, Spain, and Greece are mountainous, and dry in summer — characteristics that discourage farming. In fact, these countries are not highly productive except for regional specialties such as grapes for wine. Citrus fruits and olives, too, are important here.

On the graph shown in Figure 5-4, the vertical axis shows how great the per capita national income of a country is, while the horizontal axis shows how much food is available per person per day. Generally speaking, a poor country has a high proportion of its people engaged in farming and related industry. Such a country is poor because most of its people are working to satisfy basic

Half the Netherlands is below sea level.

When Denmark joined the Common Market in 1973, the country increased its agricultural sales by over $300 million without producing anything extra. The increase was due solely to the Common Agricultural Policy of the EC.

Marilyn Olsen

A dyke (above) shows the use of the high well-drained surface for a motor road which is several metres above the level of the new polder (below).

Marilyn Olsen

wants (such as food), and there are not many people left over to work in other industries. If you follow the area between the two curved lines, you will first pass countries that (starting with Albania and Romania) have barely adequate food supplies, produced inefficiently. Next come countries that still have relatively inefficient agriculture but sometimes manage to have food surpluses. Finally, there is a cluster of highly efficient, highly industrialized nations that produce ample food with very small farm work-forces. Such efficiency is the goal of all nations. But why is there such a variation in Europe?

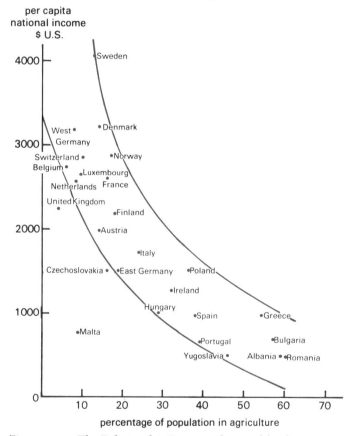

Figure 5-4.   The Relationship Between the Wealth of a Country and the Efficiency of Its Agriculture

Apart from things like climate and soil-type, the most important reason involves the wealth of each country. Since wealth is measured as *goods* and *services* (produced by people working in industry), countries with only a few people in agriculture must have many people working in industry; these countries produce a lot of goods **and**

Shell Photographic Service

Insecticides and weedkillers are applied most easily and quickly by cropdusting helicopters like the one shown above.

services and are therefore wealthy. They have many machines, a high level of technology, probably irrigation works (if required), and almost certainly ample supplies of fertilizer. Farm sizes are probably large enough, and the farms themselves well enough organized, to use all these aids most efficiently. A good measure of a country's wealth is the figure you get by dividing the value of all additional goods and services produced in a year by the total number of inhabitants of that country. This figure is known as *per capita national income*. Now that you know that, have another look at Figure 5-4. It makes sense, doesn't it?

7. Describe as carefully and completely as you can the differences between the countries in the group comprising the Netherlands, West Germany, and Sweden, and the countries in the group including Romania, Bulgaria, Greece, etc. You may only talk about the differences due to their relative levels of efficiency in agriculture, and sizes of per capita national income.

   Outline the difficulties the second group of countries would face (and have to overcome) if they tried to raise their per capita national incomes.

After all the thought involved in the answer to question 7, you should have reached the conclusion that poor countries probably have difficulty producing a food surplus because they cannot afford tractors and other farm machinery, fertilizers, and pesticides that the richer countries can (and have to). Accordingly, although the poor countries have a lot of people working on the land, it could well be a case of "too many cooks spoiling the broth." Farm sizes are too small for the land to be worked to its maximum advantage, because each farmer cannot make enough profit to afford to buy all the necessary things that a farmer with a larger land area could.

In the communist countries of eastern Europe, collective farming was introduced as a way around this problem. In Poland, for example, collective farms were introduced after the Second World War as a means of increasing the size of holdings to enable them to be worked efficiently. Although in Poland the process of collectivization only involved about 20% of the land initially, the government is having to take over more and more of the privately owned farmland because so many young people are leaving the land. Nearly 400 000 farms, involving 1.6 million ha were being worked in 1974 by people over 50 who had no heirs.

In Bulgaria, collective farms involve almost all the

In 1974, eastern Europe had bumper harvests. *Czechoslovakia* raised grain production to 10 million tonnes, an increase of 12% over 1973. *Hungary* reaped about 12 million tonnes, up 9%. *Poland* took in a little more than 2% over 1973 — well over 20 million tonnes.

farmland, as is the case with Romania and other east European countries.

A collective farm is basically an amalgamation of all the land held in, say, a village. All the people of the village hold this land in common, work on it, and share the income. Decisions over farm management are usually made by an elected board of perhaps 20 people. Tractors and other farm aids are easier to obtain and in theory, farm yields should increase, allowing many people to leave the land and go to work in industry. In practice, there seems to be a problem with human nature. People who own their own private farm feel more *incentive* to produce than those who are paid a share that doesn't depend very much on output. However, after 25 years (not really a very long time) some of the "bugs" in the farming systems in eastern Europe are being removed. In Bulgaria, higher crop

Marilyn Olsen

A farm village in Moravia. Where is Moravia?

	Yields in 100 kg/ha		Milk yield per milking cow kg/year	Arable land		Per capita national income $US
	Wheat	Potatoes		per tractor in hectares	per tonne of fertilizer (ha)	
†Albania	14.4	116	687	87	12.4	500 est
Austria	31.7	244	2956	7	4.3	1993
*Belgium	38.6	306	3755	10	1.7	2726
†Bulgaria	27.5	120	2000	86	6.6	700 est
†Czechoslovakia	28.6	156	2275	40	4.5	1500 est
*Denmark	44.2	229	3913	15	4.6	3192
Finland	21.2	152	3440	18	2.9	2178
*France	34.3	208	2982	16	4.6	2606
†Germany (East)	36.3	179	3190	33	3.2	1500 est
*Germany (West)	38.8	278	3710	6	2.7	3168
Greece	16.9	103	1110	56	11.2	998
†Hungary	24.3	103	2448	83	8.0	1000 est
*Ireland	40.8	264	2401	15	3.0	1442
*Italy	22.9	121	2081	26	12.2	1723
*Luxembourg	27.6	232	3629	8	2.7	2641
Malta	14.1	68	3657	128	33.0	745
*Netherlands	44.1	338	4208	6	1.4	2553
Norway	30.3	227	3641	10	4.5	2877
†Poland	23.1	171	2340	82	6.3	1500 est
Portugal	9.2	106	2969	173	22.1	673
†Romania	17.5	95	1651	103	18.0	500 est
Spain	12.3	120	1578	93	16.5	998
Sweden	37.8	245	3762	13	6.5	4032
Switzerland	36.8	293	3500	6	2.9	2859
*U.K.	39.5	246	3848	21	4.6	2249
†Yugoslavia	23.4	91	1233	121	12.7	500 est

*Members of the European Community (EC)
†Under communist governments
Note: Yields are averages, 1966-1970 inclusive. Other values are for the latest available
    year, usually 1970 or 1971.

Figure 5-5(a).   Some Representative Statistics for European Agriculture

yields are made by increasing the size of collective farms. In Poland, the same results are achieved by supporting the mass of private small farmers and helping them to pool local labour and farm machinery.

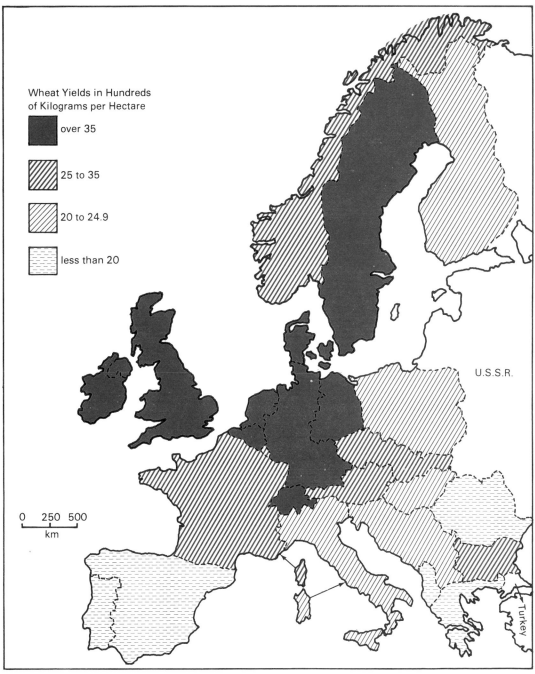

Figure 5-5 (b).   European Wheat Yields

8. All the data in Figure 5-5(a) can be portrayed in the form of maps, graphs, charts, etc. As an example, Figure 5-5(b) shows the distribution of wheat yields across Europe. Consult with your teacher about the best way to portray the data in Figure 5-5(a) and complete the representation of the data in Figure 5-5(a).

9. Write down any conclusions that you may have concerning the maps or charts that you drew for the preceding assignment. There certainly seems to be some kind of pattern, doesn't there?

Finally, to round out our study of the way farming varies across Europe, we shall examine some important crops and try to see which countries are most important in their production.

Figure 5-6 needs a little explanation. You will notice that production figures vary enormously. There are several reasons for this. A country may have the type of climate that favours the growth of certain types of crops over others. Or one country may have a much larger area of farmland than another. Or, again, there may be a much greater population in one country, so that more food has

	Corn	Sugar beets	Toma-toes	Olives	To-bacco	Grapes	Citrus fruit	Cab-bages	Pota-toes	Meat	Millions of hen eggs	Butter
Albania	240	135		14	13.5	53	3	4	118	48	64	
Austria	612	2 005	21		0.6	322		95	2 704	472	1 520	45
Belgium	9	4 400	84		1.9	12		459	1 593	605	4 000	98
Bulgaria	2 800	1 675	726		96.0	1 285		180	360	326	1 696	14
Czechoslovakia	520	5 809	147		5.8	117		330	5 500	925	3 500	87
Denmark		2 008	20					41	900	881	1 552	144
Finland		366	8					11	1 136		1 046	101
France	7 475	17 928	506	19	47.2	7 847	6	335	9 034	2 966	11 863	540
Germany (East)	3	4 856	19		6.8			256	10 500	1 052	4 364	224
Germany (West)	507	13 391	33		8.9	761		481	16 250	3 407	14 900	522
Greece	527	1 050	822	950	98.7	1 725	630	130	590	209	1 900	7
Hungary	4 743	3 301	400		27.3	939		229	1 680	439	2 800	21
Ireland		917	17					127	1 450	384	728	78
Italy	4 780	10 571	3 670	2 405	75.2	11 153	2 585	836	3 842	1 240	9 009	67
Luxembourg						17		2	58	25	78	8
Malta			6			4		1	18	6	122	
Netherlands	4	5 002	356			5		291	5 648	910	4 737	112
Norway			10					56	800		655	21
Poland	12	11 321	317		85.0			1 383	43 500	1 882	6 732	192
Portugal	591	115	730	492	13.5	1 138	142		1 201	212	655	2
Romania	7 100	3 783	676		24.0	1 189		473	2 525		3 100	27
Spain	1 859	4 980	1 398	1 765	21.5	3 946	2 262	773	4 390		7 000	6
Sweden		1 471	6					19	1 545	339	1 532	63
Switzerland	63	392	20		2.5	115		25	930	328	664	32
U.K.		6 034	94					945	6 550	1 975	14 800	60
Yugoslavia	6 961	3 636	328	7	48.4	1 499		611	3 000	596	2 604	17

Figure 5-6.   **The Production of Various Crops and Farm Products in Europe, 1970**

Note: All figures in thousands of tonnes unless otherwise stated.

to be produced. We must not forget, too, that politics has a big influence. In the countries of the European Community, governments are bound to buy everything that their farmers produce. Because the cool, damp climate of most of the EC countries particularly favours the growth of grass, dairy cattle are kept in large numbers. So much butter has been produced that it has to be "dumped" in other countries, notably in South America and the U.S.S.R. Nearly half a million tonnes of butter had been over-produced by 1968! And the governments of the EC countries felt powerless to prevent this overproduction because they would have lost the votes of the farmers!

10. Discuss with your teacher the various ways that you can show the information in Figure 5-6. Charts, maps, and graphs are all useful and interesting to draw up.
11. Which crops seem most subject to climatic controls? Give reasons with your answers.

There is no doubt about it. In a world where over half the population goes to bed hungry every night, the continent of Europe stands out as being healthy and well fed. It is an example that other parts of the world could well copy. Certainly the European countries all give foreign aid in some form or other. Whether they can encourage poorer nations in Asia and Africa to reach the same dietary level remains to be seen.

The co-operative system of farming in Europe is very well developed in Denmark. This country had once been a large wheat producer but could not meet the competition from the extensive prairies of North America when they were opened up. So farming had to change, quickly. The Danish government helped by encouraging a trend to livestock farming. New breeds of animals were developed, such as the *Landrace* pig and the *Black and White* and *Danish Red* dairy cows. Most farmers belong to a co-operative society that helps them in marketing, gives expert technical advice, runs agriculture courses, and lends money at a reasonable rate of interest. Well over half of all Denmark's farm produce goes to market through co-operative societies. It takes the form of bacon and pork; milk, cream, butter, and cheese; eggs and poultry; and beef. Most of it is exported, which is why Denmark joined the EC.

Swiss National Tourist Office

The little village of St. Stephen is typical of many Swiss farming villages. In summer while the dairy cows are higher up on the Alpine pastures, the valley floors are cultivated. One of the more important crops is hay. Why?

6. SCANDINAVIA

AREAL DIFFERENTIATION

Five countries make up this part of the European land-mass — 5 countries with a great many differences among them. The major types of difference can be summed up under the following headings:

— differences arising from the nature of the underlying rock, the shape of the land surface, and the interweaving of land and sea;
— differences in climate;
— differences in their economies.

STRUCTURE AND RELIEF

The underlying rock of Scandinavia is of many different types. Underlying Sweden and Finland are some of the oldest rocks in the world, containing some of the earliest traces of life. They date back some 3 500 000 000 years and form a hard block that has tended to resist earth movements. The rocks are very similar to those of the Canadian Shield and, especially in Sweden, contain some rich mineral lodes. Mostly the rocks are granite or of granite type.

By way of contrast, the rocks of Norway are generally much younger. They are sedimentary rocks and were laid down at the bottom of a deep sea 500 000 000 years ago. Then earth movements folded and raised them high above the level of the sea, so that the forces of erosion began to carve into them. Wind and rain, ice and frost — all contribute to the carving of the mountains of Scandinavia.

Denmark, to the south, is made up of still younger rocks. Laid down just before and during the age of the dinosaurs, Denmark's sedimentary rocks are very soft and easily eroded. Denmark is a low, flat land, including a group of islands that are easily scoured and abraded by the waves of the North and Baltic Seas.

Away to the west, across the grey, wind-tossed wastes of the North Atlantic, lies Iceland. The rocks of Iceland are very young. They lie as a collection of volcanic cones and ridges above the general level of the mid-Atlantic ridge, a vast system of submarine mountains that stretch

Millions of years ago, Scandinavia was joined with all the other regions of the world into one vast continent which slowly split apart. The parts that separated from Scandinavia were North America and the British Isles. The proof that they were once joined is given by fossils, identical on both sides of the Atlantic, and similar folded rock structures in the Kiolen Mountains, the Highlands of Scotland, and the mountains of Nova Scotia.

The first colonists of Iceland were Irish monks, seeking peace and sanctuary from the Vikings. Of course, they were quickly chased away by the Vikings to other lands. For a thousand years, the language of the Icelanders has remained unchanged. It is called Norse.

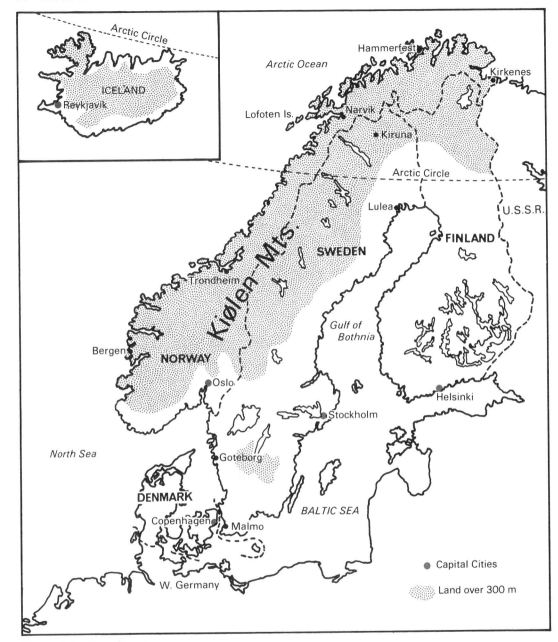

Figure 6-1. Scandinavia

the length of the Atlantic and join up with the other, similar features that are to be found in all the world's oceans. So young are the mountains and plateaus of Iceland that they are still being built by volcanic action.

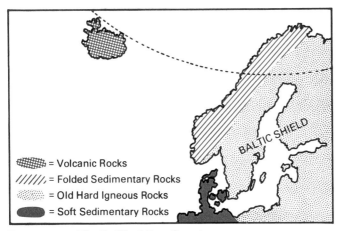

Figure 6-2. **The Build of Scandinavia**

GLACIATION

One factor common to all the different rocks and relief features of Scandinavia is that of glaciation. The last Ice Age covered the region with a great sheet of ice that carved deep U-shaped valleys where rivers once flowed, removed most of the topsoil and deposited it farther south over northern Europe, and cut the coastline into a series of *fiords*. A fiord is a deep, glaciated valley that has been flooded by the sea. There are several regions of the world where fiords are found besides Scandinavia. One is on Canada's west coast.

The fiords of Norway are particularly numerous and well developed. Although they are scenically beautiful, they do pose problems of communications. It is very difficult to travel along the coast of Norway, and ferries or other forms of sea transport are very important. As the map of the Sogne Fiord in Figure 6-3(b) shows, the fiords can penetrate a long way inland. This makes them fine harbours for fishing and trade. Most of Norway's coastal traffic takes place in the protected waters between the mainland proper and the outer island shield (the *skerries*). It is not surprising, then, that all Norway's major cities lie on fiords.

By way of contrast, the coastlines of Sweden, Finland, and Denmark are not nearly so rugged. The land is lower, and instead of there being separate tongues of ice to carve fiords, the land was smoothed down as if by a gigantic sheet of sandpaper by an all-covering sheet of ice. The region of the Baltic Sea was at the centre of the great ice

Royal Danish Embassy

These Greenland Eskimos are Danish nationals too!

The German battleship *Tirpitz* sought the cover of Alt fiord in northern Norway during the Second World War. It proved hard for Allied bombers to sink her, for the fiord walls were steep, deep, and close together, with anti-aircraft guns turning the anchorage into a death trap. In the end, though, the ship was sent to the bottom. After the war, the Norwegians salvaged the vessel for scrap.

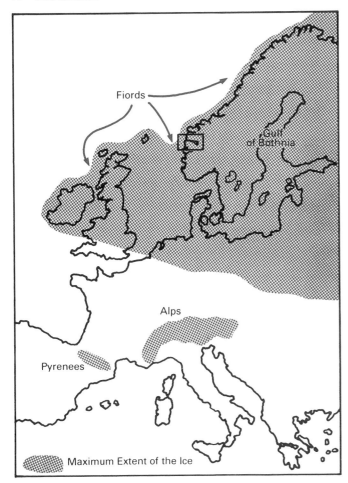

Boxed area is shown below, enlarged.

Norwegian Embassy
Information Service

Extending inland from the ocean mile after mile, Western Norway's narrow fjords provide some of the most spectacular scenery to be found anywhere in the world.

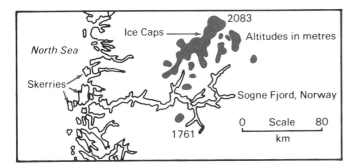

Figure 6-3 (a).   The Ice Sheets of the Last Ice Age

sheet, precisely where it was thickest. The exact depth of the ice is not possible to determine, but it certainly exceeded 1000 m. This great weight of ice actually pushed

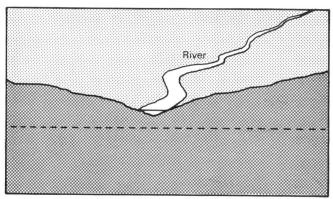

The original, Unglaciated Land Surface

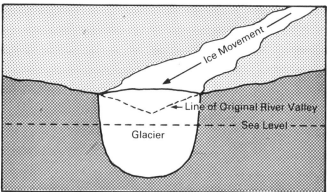

During Glaciation Ice May Erode Below Sea Level

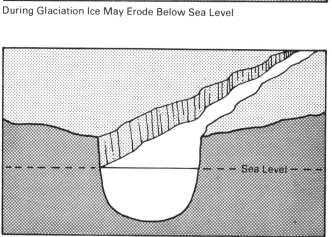

After Glaciation the Sea "Drowns" the Deep, U-shaped Valley, Often to a Depth of Hundreds of Metres

Figure 6-3 (b).   The Formation of a Fjord

Midnight sun at North Cape, Norway.

much of the land below sea level, and now that it has melted, the land is rising again, although very slowly. The south coast of Finland is rising by about 4 mm annually. The north coast of Sweden is rising at a rate of about 1 cm annually, and the whole Gulf of Bothnia area has risen about 200 m since the last Ice Age.

The remains of the great ice sheet are to be found on the high plateau region of Norway known as the *fjeld*. Even today, the winter snow cover persists for more than half the year there. Rising high above the level of the fjeld are isolated peaks, the highest of which is Galdöpiggen (2468 m).

If you study an atlas map of Sweden, you will notice a striking feature — a large number of long, narrow lakes (finger lakes) that all run roughly parallel with one another. They terminate at the start of the Baltic Shield, where the rocks are generally harder than those of the mountains. Consequently, the scooping out of the old river valleys by the glaciers to form lake hollows had to stop abruptly. Generally from the *glint line*, as it is known, the land falls gently eastwards to the Baltic Sea, to emerge in Finland as a low, well-glaciated plateau dotted with lakes that occupy over half the land area.

The ice sheet that covered the Scandinavian peninsula during the last Ice Age sent lobes of ice as far as the east coast of Britain. Glacial pebbles of Norwegian granite are commonly found there on the beaches.

The glint line is utilized by many hydro-power stations in Sweden. The power is used for homes, sawmills, and metal-refining plants.

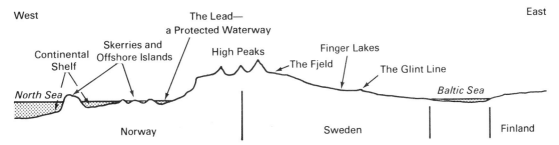

Figure 6-4.   A Cross Section through Scandinavia

To the south lies Denmark. Much of Denmark is covered by a layer of clay and loam many metres thick, deposited by meltwater streams from the retreating glaciers. This is true of the eastern part of the country. The western coastlands of Denmark tend to be sandy and gravelly and not particularly fertile, except where people have reclaimed them. Nowhere in Denmark is higher than 200 m above sea level, even though this land, too, is rising slowly after the last Ice Age.

Iceland most resembles Norway in its scenery, but there are differences. Iceland is too cool for much tree growth, and about the only reasonable soils possessed by the island are used as pasturelands. The coastline is cut deeply by spectacular fiords, which, as in Norway, are often occupied by fishing settlements.

	Total land area (hectares)	Population (1974 est.)
Denmark	4 237 000	5 000 000
Finland	30 540 000	4 750 000
Iceland	10 027 000	215 000
Norway	32 422 000	4 000 000
Sweden	41 148 000	8 200 000

Figure 6-5.   The Land Areas and Populations of the Scandinavian Countries

CLIMATE

The climate of Scandinavia exhibits some surprising contrasts in such a small area. The warm North Atlantic Drift prevents much cold Arctic water from penetrating along the Norwegian coast south of the Arctic Circle. Even in winter, sea-level coastal temperatures are above freezing well to the north of the Arctic Circle — as far as 75°N. Without the North Atlantic Drift, temperatures would be much lower in winter and the situation would be more like that on Canada's east coast, where icebergs are commonly seen at latitudes of only 50°N.

Owing to the rapid increase of the land elevation above sea level as one moves inland, the warming effect of the ocean in winter is quickly lost as the coast is left behind. The high land encourages heavy precipitation (in the form of snow in winter), because moisture-laden winds blowing in off the North Atlantic Drift quickly reach dew point as they rise over the mountains.

In summer, the coastal lands of Norway are kept cool by the same water that warms them in winter. Such regions, with cool summers and warm winters because of the sea's influence, are known as *maritime climates*. Iceland also has such a climate.

Finland, Sweden, and Denmark, on the other hand, are much more under the influence of large land areas, both in summer and winter. Thus their climates are *continental*. Land heats up and cools down 3 times faster than water. This means that their winters are much colder than those of maritime climates because land areas lose their heat very rapidly when compared with water. On the other

Swedish Information Service

Winter landscape in northern Sweden.

		J	F	M	A	M	J	J	A	S	O	N	D
Bergen	temperature °C	1	1	2	6	9	13	14	14	11	7	4	2
	precipitation mm	230	170	160	110	120	100	140	200	230	240	220	230
Stockholm		−4	−5	−3	3	9	14	17	15	10	5	0	−4
		33	28	30	30	43	51	69	71	51	53	43	41
Helsinki		−6	−7	−4	1	8	14	17	16	11	6	0	−4
		46	36	36	36	46	46	56	74	64	66	64	61
Vardó		−6	−6	−4	−1	2	6	9	9	6	2	−2	−4
		69	66	53	41	36	38	46	51	61	64	64	66

Figure 6-6. Some Climatic Data for Four Places in Scandinavia

hand, summers in a continental climate are hotter than those of a maritime type. Generally, because the sea has less influence on a continental climate, precipitation is lower. Figure 1-2 (in the chapter on climate) should make all this clearer.

It is vital to understand that climate is not just studied for its own sake. While this is important, it is just as necessary to see that climate has an influence on what can grow.

NATURAL VEGETATION

The dominant natural vegetation of Scandinavia is forest, mainly coniferous trees (softwoods). The factor that most affects tree growth here is temperature. Where it drops too low, there are no trees. In mountainous regions, while precipitation may be high, due to low temperatures, much of it may be in the form of ice and snow which plants, particularly trees, cannot use. That is why in Norway, which has a generally high precipitation, relatively little of the land area has forest cover. Sweden, with less rainfall, is not so high, and its tree cover is quite extensive. So is Finland's. Denmark was once covered with dense forest, but that was centuries ago. They have long since been cleared, because Denmark's climate is the mildest of all the Scandinavian countries, and the flat land proves suitable for farming.

Well to the north of the Arctic Circle lies a belt of cold desert or *tundra*. Little grows here, save for coarse grass and some stunted shrub-like trees in protected areas. Mosses and lichens do well, but can be little used by people. The tundra is the home of the Lapps, who live by herding reindeer. They follow their animals on their sea-

Norwegian Embassy
Information Service

Tromso lies north of the Arctic Circle in Norway but there has been settlement here for over 1000 years.

Heather Boyd

In Norway, flat land for building is in short supply.

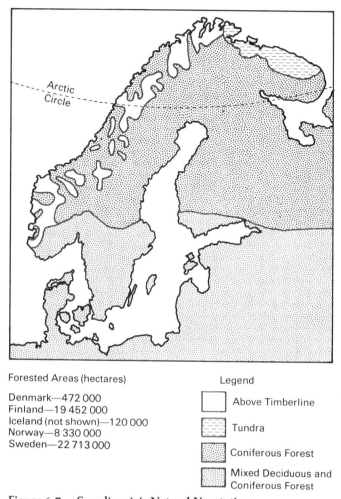

Forested Areas (hectares)

Denmark—472 000
Finland—19 452 000
Iceland (not shown)—120 000
Norway—8 330 000
Sweden—22 713 000

Legend

☐ Above Timberline

☐ Tundra

☐ Coniferous Forest

☐ Mixed Deciduous and
Coniferous Forest

Figure 6-7. Scandinavia's Natural Vegetation

sonal migrations (north in summer and south in winter), killing them for meat and hides when required, and milking the cows. Their numbers have diminished rapidly in recent years, however, and only a few thousand Lapps preserve the old ways. Most now live in cities, where the life is more secure, if less challenging in some ways. What probably destroyed their traditional way of life more than anything else was northern development in the last century, and the hardening of national boundaries.

Besides being one of the most important factors determining what plants can grow naturally, climate also determines what farmers can grow and what animals they can keep. Before we leave climate as a topic, we should see what its influences are on agriculture.

AGRICULTURE

By now you should be well aware that when the average temperature falls below 6°C in any month, the month cannot be counted as part of the *growing season*. When the growing season is less than 5 months, cultivation becomes uneconomic, if not impossible. Most crops need at least 5 months to mature. Generally in Scandinavia, south of the Arctic Circle the growing season is 5 months long, but only at sea level. Anywhere there are mountains or other high land the temperatures fall, and so the growing season shortens. A rough rule of thumb is to bear in mind that as altitude increases, temperature falls at the rate of about 0.6°C for every 100 m. So mountains 1000 m high show a temperature fall of some 6°C at the top over the bottom. That really does not leave Norway very much land with a growing season of 5 months!

Heather Boyd

A Norwegian farm.

Apart from Denmark, agricultural land in Scandinavia is very limited. *Norway* has cropland around Trondheim, Oslo, and Stavanger. The chief food crops are grain (mainly barley), soft fruits along some sheltered fiords, and vegetables grown locally near towns and villages. Most crops are *fodder* crops of hay, potatoes, and kale. They are used to feed dairy cows during the winter when grass or other natural pasture is a problem. In fact, pasture for animals in Norway is a problem in summer too, and can only be solved by taking the dairy animals high into the mountains to the fjeld. Here temporary farm accommodations house the farmers while the cows graze the high alpine pastures—called *saeters*—for 2 or 3 months. This moving about, called *transhumance,* is essential to Norway's dairy industry, and the dairy industry is the mainstay of Norwegian agriculture.

Sweden is in rather better shape as far as farming is concerned. Most cropland lies in the southern third of the country, but crops can be grown nearly everywhere in the inhabited parts of the country. As in Norway dairy cattle and their fodder requirements dominate farming life, but transhumance is unnecessary.

While Norway is extremely short of flat land, there is no such problem in Sweden. Common to both countries, though, is the problem of drying hay in the damp climate. The hay must be spread on wires strung across the fields after mowing. In Sweden the principal crops are fodder crops, and although wheat and other grains can be spring-sown as far north as the Arctic Circle, production of grains

Heather Boyd

Why do you think hay must be dried on racks like this in Norway?

Rich farmland like this is rare in Norway.

Norwegian Embassy Information Service

is not important except in the south, where vegetables are also produced for canning. Fruit orchards produce mainly apples.

While 85% of Norway's farms are less than 10 ha in size, in Sweden only 50% of farms are less than 20 ha. In *Finland*, average farm sizes are about the same as in Norway. Again, animal raising, particularly dairying, is most important. Most crops grown are fodder crops for winter feeding of the milking herds.

Iceland has little agriculture. Perhaps the most interesting development is the use of volcanic steam to heat greenhouses, which produce vegetables near the capital city of Reykjavik. Sheep are widely raised because they are so hardy and survive well on the coarse natural grass, which is often cut for hay.

Denmark, for all its small size, remains the most important farming country in Scandinavia. Like all the Scandinavian countries, Denmark uses co-operatives to make farming efficient but uses them far more intensively than the other countries. Generally, farm sizes in Denmark are about 25 ha, but larger farms are found on poorer soils in the west. The raising of animals (pigs, poultry, but especially dairy cows) is the most important activity. In fact, animals eat over 80% of everything that Danish agriculture grows.

Royal Danish Embassy

The Danish Landrace pig has been bred specially for its bacon.

	Arable land as a % of total land area	Meadows and pasture as a % of total land area	Cow milk production in tonnes (1972)	Butter production tonnes (1972)	Cheese production tonnes (1972)
Denmark	63.0	7.0	4 500 000	135 900	130 900
Finland	9.0	almost none	3 344 000	83 500	46 433
Iceland	almost none	22.0	125 000	1 430	3 727
Norway	2.5	0.5	1 780 000	21 825	56 720
Sweden	7.0	1.5	2 700 000	44 600	69 770

Figure 6-8.   Some Agricultural Production Figures for Scandinavia, 1972

There is a close association between the efficiency of farming in Denmark and the density of population. This might seem surprising, for only 14% of Denmark's labour force is engaged directly in agriculture. In Finland 28% and in Iceland 26% of the respective labour forces are engaged in farming. In Denmark, though, there are many industries dependent upon farm produce for their existence — dairies, creameries, and bacon factories, for example. In addition, the relatively high population density

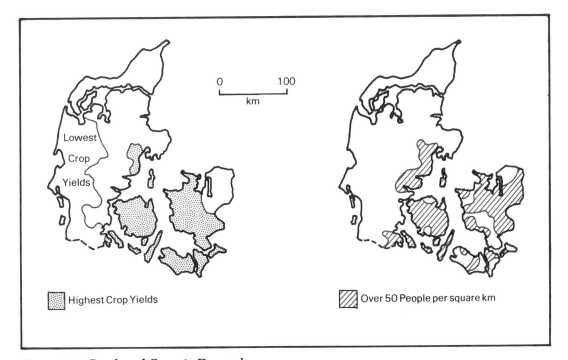

Figure 6-9.   People and Crops in Denmark

of Denmark means that there is great pressure to make the most efficient use of the land.

INDUSTRY

As with all the world's developed countries, Scandinavian countries have a wide range of industries. Nevertheless, a few industries dominate all others in the various countries; to keep things brief, we shall concentrate on those areas which appear to be national specialties.

Norway is remarkable for an old-established and still-flourishing *fishing* industry in the North Sea and the Arctic Ocean. One of the biggest fishing nations in the world, Norway occupied fifth place for tonnage landed in 1972 after the U.S.S.R., Japan, Peru, and China. (It is worth noting that all these other countries have many more millions of people than Norway does.) The chief fish caught are cod, which migrate south from the Arctic to the Lofoten Islands to breed in February and March. Farther south, off the coastline from Trondheim to Bergen, herring are caught in millions in April. The importance of fishing to Norway cannot be exaggerated. The government claims a limit of 19.2 km to its territorial waters, which it jealously guards against the fishermen from other countries.

Much of the Norwegian fish catch is exported frozen, canned or, in the case of cod, dried in the air on racks, just as was once widely done in the maritime provinces of Canada. An interesting feature is the employment of farmers in fishing. The first four months of the year are the most important months for fishing, but they are almost idle times for farmers. In many cases, where farms are

Since 1960, the number of full-time workers in Danish agriculture is down to less than 150 000 from 300 000.

The Icelandic sagas preserve the accounts of Eric the Red and Leif the Lucky, men who sighted North America and promoted expeditions there hundreds of years before Columbus. They sailed from Iceland.

Norway fishes for the world's supermarkets. Hjalmar Johansen and 2 of his sons run a small processing plant in Somaroy. They send over 2 500 000 kg of fish to the United States annually, and the entire country of Norway sends over 50 000 t of frozen fillets there in any year. The Johansens' factory benefits from the Regional Development Fund of Norway, which encourages people to live in the north. A company like the Johansens' can get back 35% of its investment funds from the government if it locates near the Arctic Circle. Established in 1961, the RDF has allowed Hammerfest, the most northerly town in the world, to grow to more than 7000 people in the period from 1962 to 1972.

really too small and remote for much profit, the farmers spend the four months at sea, supplementing their incomes while their families stay at home and do such work as is necessary on the farm.

The bulk of Norwegian fish are caught within home waters, and it was this that led to the country's deciding not to enter the Common Market in 1973. Membership in the EC would have opened up Norwegian waters to fishing by all the member states. In addition, the Norwegian government would not have been able to pursue its policy of subsidizing the northern fishing ports of Hammerfest, Tromso, and Kirkenes so that population would have streamed south of the Arctic Circle in search of other employment — which simply does not exist. While fishing only contributes about 4% to Norway's GNP, and offers only 4% of all jobs, it is a vital industry to northern Norway.

Norwegian Embassy
Information Service

Assembling drilling rigs for the Norwegian sector of the North Sea.

	Fish catch 1972 tonnes
Denmark	1 442 900
Finland	67 000
Iceland	726 500
Norway	3 162 900
Sweden	224 700

Figure 6-10. Scandinavia's Fish Catch, 1972

With so much high land at their backs, the Norwegian people have traditionally looked to the sea for a living. Starting with the Vikings, many of them earned it in a piratical sort of way, and also settled in foreign lands. Since the 19th century and the growth of world trade, many Norwegians have earned a living on merchant ships carrying freight between foreign ports. In terms of shipping tonnage per inhabitant, Norway has the largest merchant shipping fleet in the world, over 20 million gross registered tonnes of it. This works out to 5 GRT per person! A large proportion is in the form of tanker vessels.

Sweden is the most heavily industrialized Scandinavian country. At the root of its great development are the country's large supplies of high-grade iron ore, found in the rocks of the Baltic Shield. There are two main deposits. One lies in the south in a belt extending east and west in the central lakes region west of Stockholm. But iron has been mined here since the Middle Ages, so that output is not as high as formerly. Most iron ore is produced at Kiruna in the far north. Located in the tundra, Kiruna is

Norway's sector of the North Sea has yielded some important finds of oil and natural gas. The Ekofisk field near the boundary of the British sector, 350 km from Bergen, and the Frigg field contain most of the reserves, which are largely exported to the U.K. and West Germany. This is because there is a deep chasm in the sea floor close to Norway that is impossible to bridge with submarine oil pipelines. So oil sent to the U.K. is sent back to the Norwegians in tankers. Beginning in 1975, it is estimated the Norwegian government will earn nearly a billion dollars a year from oil royalties and taxes.

Swedish ice-breakers are now engaged on forging a path to Lulea that can be kept open all year. At stake is a billion-dollar investment near Lulea to produce iron and steel. It is envisaged that 7 ice-breakers will work full time during the winter months to allow the 4 000 000 t of steel that will be produced annually to be shipped out for export to the Ruhr industrial region of West Germany. Sometimes a ship gets stuck, and when the ice-breakers move in, often the sailors ski around on the ice to work out their best moves to free the vessel.

Norwegian Embassy
Information Service

A fertilizer plant on Glomfjord, just north of the Arctic Circle.

a rare example of a flourishing town where successful living is carried on north of the Arctic Circle. Most of the iron ore mined here — 99% of it — is exported; shipment takes place largely from Narvik in *Norway*! (This is because Narvik has an ice-free port all year. The Swedish port of Lulea used to be closed by Baltic ice from November to May — over half the year — but can now be reached by ice-breakers. The railway connecting all three cities is electrified and was built in the early 1890s.)

Over 32 million tonnes of ore were mined in 1973, yet Sweden's steel industry uses very little of it. Interestingly, Swedish steel production manages not to rely on fossil fuel energy supply (what are fossil fuels?), and yet is internationally competitive. This is largely because Sweden exports little crude steel. Instead, it is manufac-

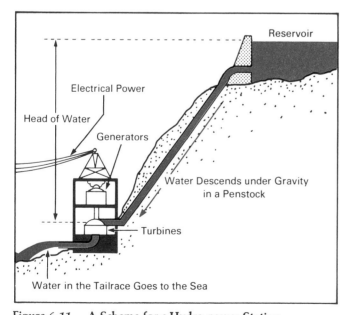

Figure 6-11.   A Scheme for a Hydro-power Station

tured into high-quality products such as ships and tankers at Göteborg, and a wide variety of goods in other parts of southern Sweden. Products such as Huskvarna sewing machines and motorcycles, SKF ball-bearings, armaments from Böfors, cars from Volvo, and planes and cars from Saab have international reputations.

The industrial power supply is based largely upon hydro-electric power. Steep slopes, fast-flowing water and high precipitation totals mean that hydro-power is relatively cheap to develop. The deep, narrow, glaciated valleys are easy to dam and are able to hold large amounts of water. An additional advantage is that electricity is relatively easy to transport to where it is needed, but can be generated just about anywhere the conditions are right.

Think for a moment about the physical geography of the Scandinavian countries. Which country is best suited to generate hydro-power? Probably Norway — that country has all the natural advantages in abundance. For this reason, Norway has found it economical to import bauxite from as far away as Jamaica and Guyana, unload it at docks that back onto aluminum smelters, produce aluminum, and then export it. There are two aluminum smelters in Sogne fiord, and one in Hardanger fiord. Both are operated on cheap hydro-power.

Swedish Information Service

During the last decade, the Swedish automotive industry has shown a rapid growth. About 45 per cent of the total production is shipped abroad.

	Denmark	Finland	Iceland	Norway	Sweden		
Electricity	17 540	20 842	1 604	62 930	66 549	total	million kW.h
	24	10 499	1 592	62 647	52 027	hydro	
	—	—	12	—	—	geothermal	
Steel	471	1 025	—	863	5 263		thousand t
Aluminum	7.5 (s)	4.7	41.0	528.6	74.4		thousand t

(s) = produced from scrap

Figure 6-12.   Energy Production and Metal Production in Scandinavia, 1971

The most important seasonal employment in Sweden is in the lumbering industry. In the winter, when the fields are covered with ice and snow, the typical Swedish smallholder pretty well has to seek employment outside of farming until spring comes. As we have seen, many join the fishing industry. But also at that time, many others go off to the forests to work in lumber camps. Most trees are cut during the winter when the bogs and rivers and lakes are frozen, making it easy to skid out logs by tractor, truck, or horse. The major method of transport for the logs is by water. Stockpiled on ice for the winter, the spring breakout assures a maximum amount of water to carry millions of logs to rivers' mouths in the Baltic Sea. There, saw mills and pulp mills create export items that are sent all over the world. (You might think that it is impossible for the different lumber companies to know which are their own logs, but in fact each company brands its logs with its own mark.)

Although Sweden produces more timber and timber products than any other Scandinavian country, it is in Finland that people rely most heavily on lumbering for a living, either directly (cutting down trees, for example) or indirectly (working in a factory making plywood, selling timber products, etc.). A typical pulp and paper plant in Finland employs over 1000 people, over 3000 in the winter. Water is essential in the paper-making process and hundreds of thousands of cubic metres are used daily. The use of water for the movement of logs in Finland is not now so important as formerly. The land is so flat and the waterways so winding that the enormous numbers of logs are most effectively moved by train and road. Only about one third of all log traffic goes by water. In Norway, very few logs travel by water but for a very different reason. There the streams are so fast and steep, broken up by waterfalls and rapids, that logs would constantly jam. Furthermore, the rivers do not have much water in them because they are so short.

Fishing in the Baltic Sea is not very important nowadays because so much DDT contaminates the water. Sprayed onto young forests in Finland and Sweden to kill insect pests, the DDT renders a lot of the fish caught dangerous to human life.

	Denmark	Finland	Iceland	Norway	Sweden	
Lumber	395	7 350	—	2 120	12 447	thousand m³
Wood pulp						
Mechanical	—	2 008	—	1 244	1 448	thousand t
Chemical	74	4 238	—	904	6 386	''
Newsprint	—	1 306	—	554	368	''
Other paper						
and board	293	3 117	—	863	3 270	''

Figure 6-13. Wood Products Production in Scandinavia

It is easier to understand how important lumbering is in the Scandinavian countries if you realize that Norway has laws regulating the use of a farmer's woodlot. He cannot cut down his own trees whenever and however he wants. The government relies on the lumber trade for a lot of foreign exchange, and trees cut down in any number must be replaced with saplings or seedlings to ensure a continuity of lumber supply.

Trees in Finland are known as ''Green Gold.''

Iceland is very dependent on fishing. Although its catch is not particularly large, the actual weight of fish caught per inhabitant is the highest in the world. Fish products make up most of Iceland's exports, by which the country earns foreign exchange. In this way Iceland can buy articles such as machinery, chemicals, and steel on the world's markets. Without fish exports, Iceland would have to look elsewhere for foreign exchange, but there is no clear alternative to the fishing industry. For this reason, Iceland has recently moved to protect its offshore fishing grounds and the cod-breeding areas in international waters by extending its territorial waters, even though this has resulted in quite a lot of complaint from other fishing nations. But, as Iceland says, if the fishing grounds are not protected, then they will inevitably be ruined. What is everyone's property is no-one's responsibility. Foreign vessels would still be allowed to fish there. It is just that quotas would be enforced so that a sustainable yield would result.

Most people in Iceland live in coastal settlements, making their living from fishing. The interior is too barren and bleak for much farming. Even though there are many active volcanoes, the land still has ice-cap remnants from the last Ice Age. Hot springs and geysers are found in many areas. In the capital city of Reykjavik, volcanic steam is even used to heat homes and apartments. The climate is relatively mild, considering the high latitudes, and it is rare for the coast to be ice-bound. The last time this happened was in the early 1970s on the northern part of the island. Then a large number of polar bears appeared in search of food. They had come south from Greenland.

Although Iceland has few natural resources that can be developed, its human resources have been harnessed effectively. There are two Icelandic airlines that link up with Europe and North America. There is a NATO base. Tourism is encouraged, and there is a flourishing industry in fishing-boat repair. The North Atlantic can be very rough in the winter and many foreign fishing vessels find it economical to put into Icelandic ports in order to effect repairs so that they may return to fishing as soon as possible.

Denmark, as we have already noted, is heavily involved in agriculture. But in terms of exports agricultural products have declined a lot in recent years. In the early 1960s, the export of food and food products accounted for nearly half of all Denmark's exports. By the early 1970s, the

Royal Danish Embassy

Danish teak furniture earns money all over the world.

In the winter of 1973-1974, the average homeowner in Reykjavik spent $170 on his heating bill. A house heated geothermally has no boiler, no chimney, and no pollution.

proportion was down to one-third. There was no drop in the amount of food exports, though. Rather, there has been a much faster increase in the rate of export of manufactured goods. Tankers of hundreds of thousands of tonnes are built in Danish shipyards, and steel is even exported, although the country has no reserves of iron ore or reserves of fossil fuel. All these raw materials have to be imported. There is no hope of developing hydro-electric power, of course. The country is too flat.

	Denmark	Finland	Iceland	Norway	Sweden	(Canada)
Tractors	175 500	160 500	9 900	93 112	172 000	(638 545)
Fertilizer	307 135	182 108	13 313	81 500	235 576	(336 000) t

Note: For purposes of comparison, Canada's 1972 population was 22 000 000 and the cropland area was 43 767 000 hectares.

Figure 6-14.   Two Agricultural Aids in Scandinavia, 1972

This, then, is the Scandinavian region: glittering lakes and snow-capped peaks, empty tundra and endless forests, modern cities and centuries-old wooden churches, hot summer days in places, a shining sun at midnight. It is a region of contrasts, not only within the region, but between it and other regions, too. Scandinavia is unique.

QUESTIONS AND EXERCISES

1. What is the density of population in each of the Scandinavian countries? How do these densities compare with Canada's population density figure of one person to every 45 ha and a figure for the Netherlands of 0.07 ha per person?
2. In the exercises at the end of the chapter in this book dealing with climate, you are shown how to draw hythergraphs. Construct a hythergraph to show the climates of the cities in Figure 6-6. Describe the effect of the North Atlantic Drift on these cities once you have located them in the atlas.
3. Construct a series of percentage bar graphs to show land use in the Scandinavian countries. Divide the bars according to *cropland, forest,* and *"other"*.
4. Which country in Scandinavia produces the most dairy products per capita? Use the data in Figure 6-8 to help you, and show your answer by means of a graph.
5. Why would Scandinavian countries find it hard to produce hydro-electricity at a certain season of the year? What is the season?
6. Describe your life as a farmer in Norway, somewhere near Trondheim.
7. Compare the Scandinavian countries' agricultural characteristics shown in Figure 6-14. Use graphs to illustrate your written answer.

7. SWITZERLAND

SUPPORT OF LIFE

Switzerland is a fairly small country, with an area of just a little over 41 000 km². This is less than half the size of Portugal or Hungary, and about one-seventh the size of Italy; or, in Canadian terms, about three-fourths the size of Nova Scotia. In comparison with Asian countries, it is slightly smaller than Bhutan (find it on an atlas map) and slightly larger than Taiwan.

It holds 6 500 000 people. The density of population is therefore about 160 people per square kilometre, which is not far short of India's density of about 170 people per square kilometre. However, Switzerland's people live at a vastly different level from that of most people in India; indeed, at a vastly different level from that of most people in the world. Measured very simply by per capita Gross Domestic Product (the total money value of all the things produced inside a country in a year, divided by the popu-

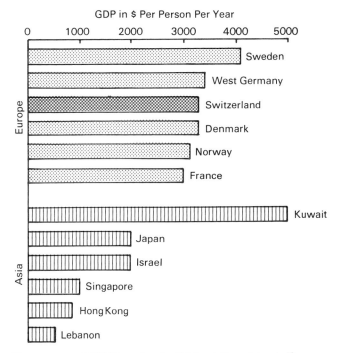

Figure 7-1. **GDP Data for the Richest Countries in Europe and in Asia**

lation of the country), Switzerland outranks all other countries in Europe except Sweden, Denmark, and West Germany. And it greatly outranks *all* the countries of Asia except some of the newly rich Arab oil states.

What does Switzerland have that enables the people to live so well (and so numerously, considering the size of the country)?

THE NATURE OF THE LAND

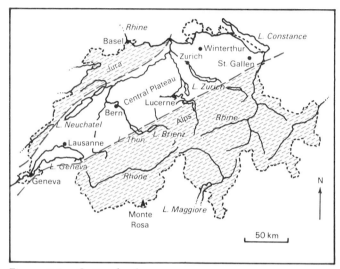

Figure 7-2.   Switzerland

Swiss National Tourist Office

The Bernese Alps near the Jungfrau.

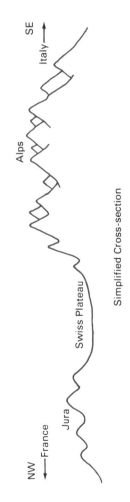

The three divisions shown in the map are all quite different from one another. In the northwest the Jura Mountains consist of parallel ridges of limestone aligned in a southwest-northeast direction. In between the limestone ridges lies a series of clay valleys. Heights rise to about 1500 m.

Across the centre of Switzerland, from Lake Geneva in the southwest to Lake Constance in the northeast, lies the Swiss Plateau. It is lowest, about 500 m, near the Jura Mountains and the northern borders along the Rhine west of Lake Constance; but it rises to about 1500 m along the edge of the Alps. The surface of the plateau is by no means flat; it just *appears* to be flat compared with the mountains to northwest (Jura) and southeast (Alps). In fact, the plateau surface is very hilly. There are lots of small ridges and isolated hills, and the rivers often flow in deeply cut valleys, sometimes even in gorges. The numerous hollows that used to occupy the surface have now been filled with rainwater, forming many lakes, some of them quite large (Lakes Geneva, Lucerne, Neuchatel, Thun, Brienz, Zurich, Constance). The rocks of the surface of the Swiss Plateau are chiefly sandstones, covered in many places by deep layers of glacial sands and clays.

The Alps occupy the entire southeastern half of the country. They are a very complex example of fold mountains, trending in the same general direction as the Jura (southwest-northeast). The rocks are a great mixture of different types, but granites and limestones are very common. They have been thoroughly *dissected* by rivers and glaciers, so that there are now many valleys cutting through the Alps. Most of the main valleys follow the general southwest-northeast trend of the rocks (for example, the upper Rhine and the upper Rhone), but there are also many important — though smaller — valleys cut-

Swiss National Tourist Office

View of Les Charbonnières in the Jura.

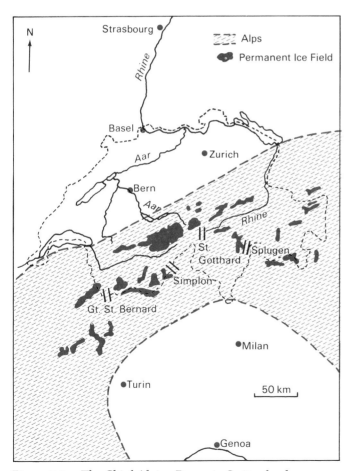

Figure 7-3. **The Chief Alpine Passes in Switzerland**

Swiss National Tourist Office

Farmland on the Swiss Plateau.

ting the mountains in a northwest-southeast direction. These smaller tributary valleys have been cut by rivers and glaciers descending from the central ridges of the Alps towards the lower lands of the Swiss Plateau to the northwest and the North Italian Plain to the south.

As a result of all this dissection, the Alps are not very difficult to cross. For centuries people have used various passes through these mountains. Even so, there are still parts of the Alps that are almost inaccessible except by really determined climbers, and large areas still support permanent icefields and glaciers. In fact, most heights rise to between 3000 and 4500 m or so, the highest point actually being Monte Rosa at 4634 m (see if you can find it in the atlas). The permanent icefields and glaciers are now confined to the higher areas, but at one time — more than 10 000 years ago, during the Ice Age — they pushed their way out of the high parts and gouged their way down the already existing river valleys. Some of these valleys now contain deep lakes, where the ice excavated hollows, or where the ice dumped its erosion debris across a valley, forming a giant natural dam. Examples of these lakes include lakes Lugano and Maggiore.

THE WEATHER

The weather is what you would expect in a mid-latitude (between 46° and 47°N) mountainous country in the interior of a continent.

What *do* you expect? You should be able to put some general principles to work. To start with, which winds prevail in these latitudes? If you said Westerlies, correct. Where do these winds come from? They come from over the Atlantic Ocean, as you know. Are they likely to be moist or dry? As you say, they will probably be moist. Are there any mountains between the Atlantic Ocean and Switzerland? Your atlas tells you that there aren't; there is only the Central Plateau of southern France, and it is not so high as the Alps, is it? So does some of the Atlantic moisture reach Switzerland? It does, and when the westerlies are forced to rise over the Jura and the Alps, what do you think happens to the moisture? Correct again — there is a lot of precipitation. But — as you note — we cannot say whether it is rain or snow, because we still have one factor missing — temperature.

What do you think the temperatures should be in a mid-latitude mountainous country in the interior of a

Prevailing Westerlies

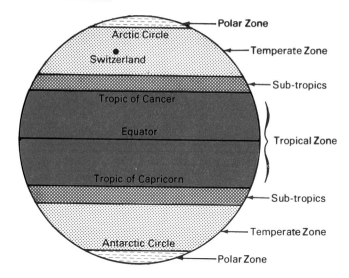

continent? Mid-latitude makes them what we call *temperate*, and mountainous country is likely to make them colder than normal for the latitude. In fact, temperatures *are* in the *temperate* range, being generally in the zone of 0°C to 25°C, and they *are* slightly colder than in the surrounding lowland areas. For example, compare the data in Figure 7-4. You will notice that Lyons, in lowland France to the west, has higher temperatures in both summer and winter than any place in Switzerland. You will also see that, within Switzerland itself, temperatures are generally higher in the plateau than they are in the Alps.

		J	F	M	A	M	J	J	A	S	O	N	D	
Nantes	a	4	5	7	10	13	16	19	18	15	12	8	4	range: 15
alt 37m	b	76	61	63	66	58	56	48	48	51	81	86	84	total: 778
Lyons	a	3	6	9	12	15	18	21	21	17	13	9	4	range: 18
alt 196m	b	33	43	58	61	79	81	66	89	76	104	79	56	total: 825
Geneva	a	1	2	6	10	14	18	20	20	16	11	6	.2	range: 19
alt 405m	b	41	51	51	64	76	76	84	85	91	122	83	56	total: 880
Lucerne	a	−2	1	5	9	13	17	19	18	13	9	5	−1	range: 21
alt 498m	b	36	51	61	91	122	145	165	143	135	92	61	50	total: 1152
St. Gotthard	a	−6	−4	−2	2	6	11	13	12	7	2	−2	−5	range: 19
alt 2096m	b	150	137	200	206	203	175	185	190	198	247	206	185	total: 2282
Jungfraujoch	a	−15	−14	−11	−9	−6	−3	0	0	−2	−6	−10	−14	range: 15
alt 3578m	b	140	183	165	206	193	241	305	267	211	178	122	150	total: 2361

a: temperature °C
b: precipitation mm (rainfall equivalent if precipitation is snow)

Figure 7-4.   Selected Climate Data

Switzerland's location, in the interior of a continent, will also have some effect on the temperatures. Can you see what effect that is just from the data in Figure 7-4?

With what we have just deduced about temperatures, we can predict what form precipitation will take. It will be mostly rain in summer and fall, except in the high mountains, where it will be snow. In winter it is likely to be snow in most places (and pretty well everywhere in January). Can you now see why the high mountains are the home of permanent icefields and glaciers?

There are some other aspects of the Swiss weather that you might not be able to figure out for yourself. One of the most important things is the variability of the weather. Westerlies normally bring variable weather, but mountains exaggerate the variability. Swiss weather changes very quickly and very frequently. It can often be very sunny in the morning and become a raging thunderstorm by lunchtime, perhaps continuing rainy or maybe turning sunny again in the afternoon. Who knows? Maybe it will just rain all day. The weather forecasters certainly have a hard job. Another special characteristic of the weather in winter is the occasional bitter-cold north easterly wind called the *Bise*. It comes from the cold air over Germany, which itself often comes from Siberia in winter, and it is funnelled across the Swiss Plateau between the Jura and the Alps (see Figure 7-5).

There is also the *Föhn* wind, part of the normal Westerly Wind set-up. What happens is that the Westerlies rise over the western slopes of the Alps first, and in doing so drop some of their moisture. When the winds descend on the eastern slopes of the mountains, they become warmer. They have to lose some of their moisture on the western slopes because they get *colder* as they are forced upwards; but this process of condensation causes the release of the *latent heat* (heat which was used in evaporating the water from the Atlantic Ocean in the first place). When the winds descend on the eastern slopes, they effectively bring this latent heat with them, and so they come down *warmer than when they went up*. This apparent miracle is highly prized by the Swiss farmers because the warm winds, especially in springtime, help to melt the snow and start the crops growing quickly.

Swiss National Tourist Office

The Jungfraujoch icefield in August. The glacier moving from it is the Aletsch, the largest in Europe.

MINERAL WEALTH

Switzerland is not an important mining country. Some

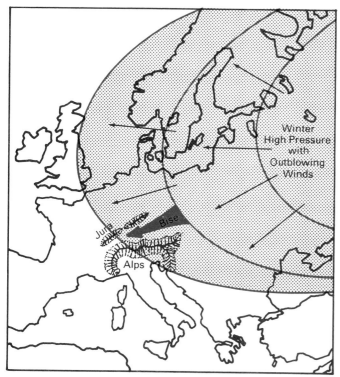

Figure 7-5. Climatic Conditions for the *Bise*

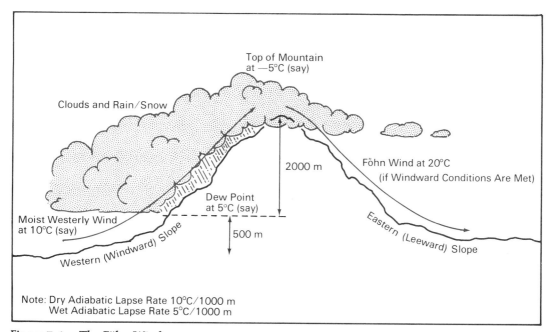

Figure 7-6. The Föhn Wind

salt mining is carried on, but not much — a little over 250 000 t a year. Iron ore used to be mined as well, but in tiny quantities, and the mines are now closed (the last one closed in 1967).

HOW THE SWISS COPE

You certainly could not say that the land we have described is one that offers an easy life; nor one that is potentially flowing with milk and honey. Yet the Swiss have nevertheless achieved one of the highest living standards in the world. How have they done it? First of all — and perhaps most important — they have made the maximum use of every little advantage that their environment offers. And second, tied in with the first point, they have worked extraordinarily hard.

One of the main advantages that their environment offers is the number of passes through the mountains, connecting the Mediterranean lands in the south with the Germanic lands in the north. Elsewhere the mountains are a fairly effective barrier, so the availability of passes in Switzerland is a distinct advantage. The passes have been used for centuries, long before the Romans. Yet it was the Romans who first stationed men permanently in the Alps to guard the passes. Later, it was the guardians of the passes who rebelled against outside control, and who set up the first beginnings of the Swiss nation (in 1291). The strategic location across some of the main north-south trade routes in Europe was therefore a very early element in the support of the Swiss nation. Because of this ability to control north-south trade, the Swiss eventually sought a policy of peace, because trade does not flow during wars. The more peaceful were the surrounding lands, the better the Swiss liked it. The Swiss officially adopted a policy of neutrality towards trouble in 1515. Three hundred years later, in 1815, the other countries of Europe recognized Switzerland's desires to remain peaceful.

Trade was not only profitable to the Swiss, it was essential. In order to eat, the Swiss have to import food. Only about 6% of the land is under crops, chiefly wheat (half the cultivated area) and potatoes. There are also smaller quantities of sugar beet, rye, barley, vegetables, and tobacco. Most of these products are grown on the Plateau. As well as crops, the Swiss also produce some fruit. For the most part the fruit is typically cool-temperate — apples and pears chiefly — but there are also some quite impor-

Try to find out the story of Hannibal and the elephants.

tant grape-growing areas (rather surprisingly for such a mountainous country). The main fruit areas are all on south-facing slopes, in order to trap the maximum amount of sunshine and get the maximum amount of heat from it. Figure 7-7 illustrates the principle.

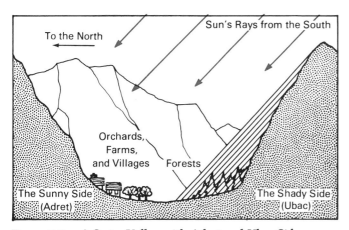

Figure 7-7. A Swiss Valley with Adret and Ubac Sides

The Swiss use a lot of their farmland for pasturing cattle. About 40% of the land is used this way. The regular supplies of water from the rain and melting snow create conditions more suited to dairying than beef ranching, and for many years Switzerland had surpluses of fresh milk. In order to conserve the milk for trading (and for use during winter) the Swiss developed skills in cheese making. Exports of Swiss cheese, chiefly Gruyère and Emmentaler, are now quite important. The importance of dairying was also enhanced by the Nestlé firm, headquartered at Vevey near Lausanne, which pioneered the manufacture of condensed milk and later branched out into chocolate manufacture. Most of the cattle — about 2 million altogether — are kept on Plateau farms, where they are usually stall-fed on silage, but some are also kept in the Jura and the Alps. In the mountains the cattle are stall-fed in winter, but in summer they are taken up into the mountain pastures (called *alps*) and allowed to graze in the open. In fall they are rounded up and brought down again, usually with a traditional folk ceremony, for it is a sign that winter is near. It is the cattle on the mountainous *alp* pastures that wear the well-known cowbells that ring whenever the animals move. Can you think why? This style of farming, where cattle are taken to *alp* pastures in summer and brought down to stalls in winter, is,

Swiss National Tourist Office
The village of Andermatt, located outside the area of shadow.

Swiss National Tourist Office
Swiss cheese being prepared for export.

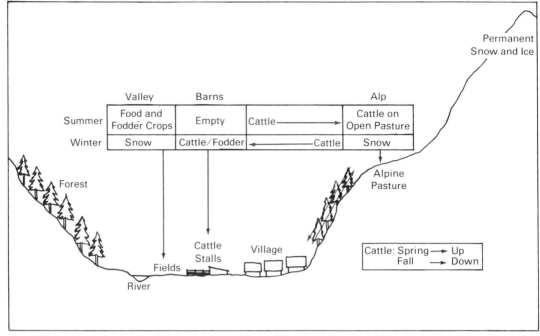

Figure 7-8.   The Transhumance System

as you know, called *transhumance*. Figure 7-8 illustrates the idea. (Where else is transhumance practised?)

As you can see, farming in Switzerland is quite varied. But it cannot produce nearly enough of most things to feed the Swiss people. Even the silage for the cattle has to be brought in from elsewhere. True, there are exportable surpluses of some things, such as cheese, but there are deficits for most things, and these have to be covered by imports. You can probably see that the Swiss interest in trade will help here, but the obvious question is, *how do the Swiss pay for the imports?* They do it in two main ways. Let's look at them more closely. . . .

MANUFACTURING INDUSTRY

Despite the lack of minerals, the Swiss have certain advantages for industry. The main one is the long-established links with other countries via trading connections. The second advantage is the combination of mountains and guaranteed water supplies, yielding a large and important hydro-electric potential, which the Swiss have almost fully developed. The chief power sites are either in the Alps or in the valleys of the large rivers crossing the Plateau (especially the Rhine, the Aar, and the Reuss

Proportion of Switzerland's
Labour Force Employed in

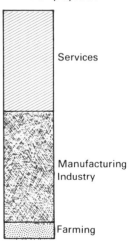

— look them up in an atlas). Hydro production now accounts for about 90% of all electricity output in Switzerland.

But the Swiss have virtually no minerals. What are they to do? Obviously they must import them. Nevertheless, importing minerals is quite expensive, especially as Switzerland is land-locked. Importers cannot therefore take advantage of cheap ocean transport; the best they can do is import by barge up the Rhine. The Swiss port that handles this trade is Basel (or Basle, if you prefer the French version). As a consequence of the relatively high cost of importing even such ordinary minerals as iron ore, the Swiss try to make the maximum use out of it once they have imported it. In this way they can make a better profit on the deal. It is clearly not going to pay the Swiss just to make bars or slabs of steel. Anyone can do that much more cheaply than the Swiss can. What the Swiss do is concentrate their production on such complicated and relatively expensive items as electric generators, turbine blades, watches, locomotives, and hi-fi equipment. Most of this *precision engineering* is located in the Zurich district, but all the major towns have some. Watchmaking is the least Zurich-oriented of the engineering industries; it is located mostly in the southwest, in Geneva and along the foot of the Jura.

Not all Swiss businesses import minerals. Some of them import textile raw materials, especially cotton and silk, which they manufacture into high quality clothing, sheets, draperies, etc. The *textile industry* is, like engineering, concentrated in the Zurich area, with a very important sub-centre at St. Gallen (find it!).

Still other businesses import chemical raw materials, chiefly from Germany via the Rhine. The chemicals are then used to form the basis of a variety of industries, some of them, such as dyestuffs, serving existing established industries, and others, such as pharmaceuticals, standing as important industries in their own right. The *chemical industries* are located just where you would expect them to be: at Basel, which is Switzerland's Rhine port, and in the northeast industrial area, chiefly Zurich, where most of the established user-industries are located.

With these three important branches of manufacturing industry — precision engineering (including watchmaking), textiles, chemicals — Switzerland is well set towards helping its people pay for at least some of the food imports they need, especially as the products of all three

90% of Switzerland's electricity is generated by hydro means; the remaining 10% is mostly thermal. What else could there be?

The Swiss export somewhere between 50 million and 60 million watches each year, enough to provide one for everyone in the world in about 75 years — except, of course, that in 75 years' time there will probably be 4 or 5 times as many people in the world as there are now. Do you think the world will *ever* have enough watches?

Hydro dam and reservoir in the Grimsel Pass.

industries are designed to be very valuable in relation to
their bulk. This doesn't mean that all the products are
small and valuable; after all, electric generators for hydro-
power plants are not particularly small. It does mean,
however, that they are all individually valuable *for their
size*. A small watch costs a lot less than a giant turbine
blade, but both are valuable *for their size*. This quality
makes it relatively easy for the Swiss to export their
manufactured goods, because the costs of transportation
form only a minor part of the total costs of the items.
For example, if a turbine blade costs $250 000, and it is
well-engineered, no one is going to object very much to
an extra $1000 or so for transporting it across a continent.
Or if you would pay $100 for a good Swiss watch, you
would not mind if that price included $1 for shipping.
The concentration on quality, then, plus the high value in
relation to size, generally means that the Swiss manage
successfully to export their manufactured goods.

Even so, manufactured exports are not enough to pay
for all the imports — raw materials as well as food.
Figure 7-9 shows that the gap between exports and imports is a permanent thing. But there is another way in
which the Swiss help to pay for their food imports (and
close the trade gap at the same time, of course).

SERVICE INDUSTRIES

As well as *making* things for other people, the Swiss also
do things for other people. They get paid for it, of course.
These *doing* jobs are called service industries.

The chief example in Switzerland is banking. The
Swiss look after other people's money for them, and they
do it very safely and very quietly. They then invest it in
all sorts of profitable enterprises around the world. The
Swiss make a lot of money by doing this — enough, in
fact, to more than half-close the trade gap. Swiss banks
have gained a good reputation around the world, chiefly
because of the neutrality of the country (more or less
guaranteed since 1815, remember), the wide and knowl-
edgeable trading links acquired by the Swiss, and the determination of the government to maintain stability and
security — as well as *safety* and *discretion* (it is very difficult to find out who has deposited how much money with
a Swiss bank). Allied with banking and investment is the
insurance industry, which has prospered for much the

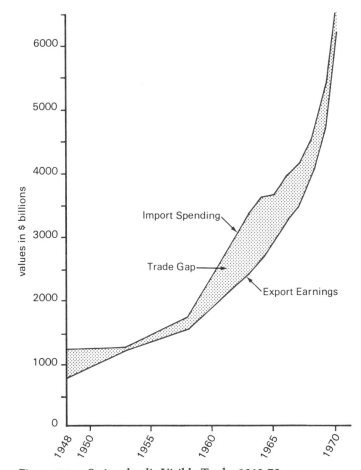

Figure 7-9. Switzerland's Visible Trade, 1948-70

Swiss National Tourist Office

Mountaineering above the snow.

same reasons as banking. You can probably name at least one important Swiss insurance company, can't you?

The tourist industry helps to close the trade gap still further. About 100 years ago people began to get interested in the Alps as something to climb, a new sort of challenge. From that, interest gradually spread until people were quite prepared to come to Switzerland just to look at the mountains, without ever wanting to climb them. Still other people, chiefly from prosperous industrial areas elsewhere in Europe, came and marvelled at the clear air in the mountains. So the health aspect of the tourist industry was born. Later still, there were people who thought that the snow in the mountains could be used for sliding around on, and so winter sports activities became a major part of the tourist industry. That is about where the tourist business stands now. There are climbers,

There are Swiss National Tourist Offices in
Argentina (Buenos Aires)
Austria (Vienna)
Belgium (Brussels)
Canada (Toronto)
Denmark (Copenhagen)
Egypt (Cairo)
England (London)
France (Paris and Nice)
Germany (Frankfurt and
  Dusseldorf)
Italy (Milan and Rome)
Netherlands (Amsterdam)
Spain (Madrid)
Sweden (Stockholm)
U.S.A. (New York, Chicago,
  and San Francisco)
In Toronto the address is Commerce Court West, Suite 2015, Toronto, Ontario.

viewers, invalids, and skiers; who knows what will come next? There is a vast amount of money earned from the tourist industry — so much, indeed, that the Swiss government has opened offices all around the world to try to encourage visitors to come to Switzerland for a vacation.

There are other service industries apart from investment banking and the tourist trade, but nothing of similar importance. One of the more interesting smaller activities is the provision of carrying services for other nations to use. Just as Norway sells the use of its merchant fleet, Switzerland tries to sell the use of its national airline (which is called . . . ?). Another small-scale activity is that of catering to international organizations. Sometimes the international groups come to Switzerland (usually Geneva) just to hold talks, but more usually they are headquartered in Switzerland permanently. Examples include the International Red Cross (which was actually founded by a Swiss), the World Health Organization, the International Labour Organization (all in Geneva), and the Universal Postal Union (in Bern — or Berne, the French version). Switzerland's guaranteed policy of neutrality and peace is obviously a very important factor in the location of these international organizations.

CONCLUSION

There is no doubt that the Swiss lead a prosperous and peaceful life. The reasons are quite varied: hard work and determination, efficient (though limited) farming, concentration on high-quality manufactured exports, and full preparedness to provide a wide range of services. By these means the Swiss are able to lead a more prosperous life than the people of many other nations.

Switzerland is a multilingual country; it has 4 national languages. Among Swiss nationals 74.5% speak German, 20.1% speak French, 4% speak Italian, 1% speak Romansch, and 0.4% speak something else. Switzerland is therefore officially known as Schweiz, Suisse, or Svizzera.

The Swiss flag is a white cross on a red background. What do you think the flag of the International Red Cross looks like?

Swiss National Tourist Office
Winter sports are possible from November through April.

QUESTIONS AND EXERCISES

1. Find out what you can about *invisible exports*.
2. Attempt to discover what the geological structure of the Alps is really like.
3. Research into the *canton* system of government.
4. Switzerland has 4 official languages. Why is this? What are they? Are there any advantages that you can think of? Disadvantages?
5. The railways in Switzerland are electrified, as they are in much of Europe. What are the advantages of electrification in Switzerland? Can you think of various advantages and disadvantages for road transport as well as rail transport inside Switzerland?

8. LONDON, ENGLAND

AREAL DIFFERENTIATION

British Information Services

Closely packed high rise buildings, old and new.

Cities everywhere in the world have at least one thing in common: they are all different from the surrounding countryside. The differences occur in a number of ways. For example, cities are usually more densely populated than the surrounding countryside. They also usually look different, with more closely packed buildings, more high-rise buildings, more public buildings (such as churches, libraries, hospitals, and so on), and more traffic. They appear to be much more crowded, with thousands of people lining up for buses, crossing roads, shopping in the stores, and just plain walking along the streets. Cities also *do* different things, compared with the countryside. They manufacture goods and provide services to a far greater degree than the countryside does.

Greater London	4200 people per square kilometre
Metro Toronto	4150
Hong Kong	4000
Singapore	3200
New York Metropolitan Area	1000
Calcutta Metropolitan District	6750

However, even though they are different from the countryside, that does not mean that they are uniform areas. There is, indeed, a great deal of *areal differentiation* within a city; one part of a city may be as different from another part of the same city as either of them might be from the surrounding countryside. For example, the factories and offices and stores and transportation routes and houses and apartments and theatres and so on do not all occupy the same areas. They occupy different areas within the city. And the people in the city *do* different things in different parts of the city. Indeed, the people themselves may even *be* different in different areas of the city.

Figure 8-1 illustrates one of the longest-held ideas about areal differentiation within a city (any city — see how it applies to your own). Obviously not all cities are circular in shape, nor even close to circular, nor do they all have exactly five major zones. Nevertheless, there is general agreement among geographers that many cities do come at

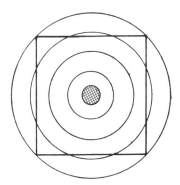

application of the
Concentric Theory
to a square town

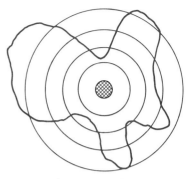

application of the
Concentric Theory
to a star-shaped town

Zone 1   Central Business District (CBD)
Zone 2   Old Industrial and Residential Areas
Zone 3   Old Residential Areas
Zone 4   New Suburban Industrial and Residential Areas
Zone 5   New Outer Residential Commuter Areas

Figure 8-1.   The Concentric Theory of Urban Structure

least fairly close to this theoretical model. The Concentric
Theory of urban structure, by the way, was first developed
by some Chicago geographers in the 1920s. Even though
several people have suggested alternative theories the
Concentric Theory still has a lot of support. How do you
feel about it, now that you have tested some of the towns
you know against it? After all, any theory should be continually tested against the reality of a lot of different cases.
If it is a good theory, then it will pass these tests, because
the real examples will support the theory (in all essentials,
at least). Let's see how the real example of London supports the theory.

In London, the downtown district (called in geographi-
cal terms the *Central Business District* or CBD) occupies
part of the north side of the River Thames. It is sur-
rounded by old Zone 2 industrial and residential areas,
especially towards the docks, which are also mostly
on the north side of the River Thames. The Zone 3 resi-

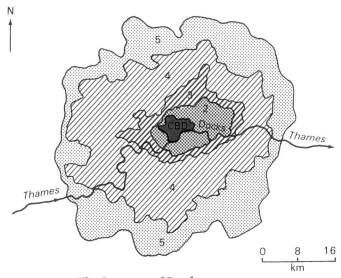

Figure 8-2. The Structure of London

dential areas extend all around the first 2 zones, particularly to the north and west, away from the older Zone 2 industrial areas of what is called the *East End* of London, near the docks. The Zone 4 industrial and residential area surrounds Zone 3, again especially to the north and west, but also this time to the south as well, where land was cheap when the buildings were being put up. Finally, the Zone 5 outer residential commuter areas stretch into the chalk hills that edge the wide Thames Valley at London (see Figure 8-3). Beyond Zone 5 lies the Green Belt. (What is a Green Belt?)

Taking a closer look at London, we can see that there is a lot of areal differentiation even within the single zones. The Central Business District, for example (see Figure 8-4 as an indication of the extent of the business concentration in central London), is not uniform throughout its whole extent. The reason, of course, is basically that there are so many different types of business. Like birds of a feather, businesses also tend to flock together, so that financial businesses cluster together, consumer businesses locate together, and so on. All the financial businesses (mainly banks and insurance companies, plus the stock exchange) are situated fairly close to one another in what is called *The City* (see Figure 8-5). The consumer businesses (chiefly stores and theatres) are generally concentrated in the district called the *West End*. The administrative businesses (parliament, civil service, the

In town-planning terms, a Green Belt is one thing and Green Wedges are another. What are Green Wedges?

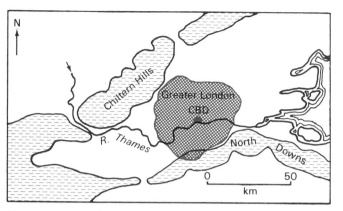

Marilyn Olsen

Piccadilly Circus: the centre of the consumer business area.

Figure 8-3.   London in Relation to the Thames Valley

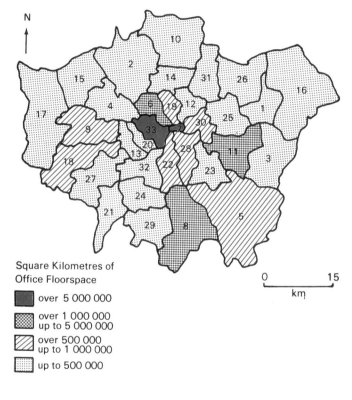

Square Kilometres of
Office Floorspace

■ over 5 000 000

▨ over 1 000 000
up to 5 000 000

▧ over 500 000
up to 1 000 000

□ up to 500 000

1	Barking	13	Hammersmith	23	Lewisham
2	Barnet	14	Haringey	24	Merton
3	Bexley	15	Harrow	25	Newham
4	Brent	16	Havering	26	Redbridge
5	Bromley	17	Hillingdon	27	Richmond
6	Camden	18	Hounslow	28	Southwark
7	City of London	19	Islington	29	Sutton
8	Croydon	20	Kensington and	30	Tower Hamlets
9	Ealing		Chelsea	31	Waltham Forest
10	Enfield	21	Kingston	32	Wandsworth
11	Greenwich	22	Lambeth	33	Westminster
12	Hackney				

**Figure 8-4.   Office Floorspace in the London Boroughs, as an
Indicator of the Central Business District**

monarchy) are all located in *Westminster*, where the name of the main street — Whitehall — has now come to be used as another name for the government itself. Indeed, this tendency is characteristic of London. Areal specialization is so well developed that even just the name of a single street can indicate the most important activity on that street. For example, Oxford Street now indicates large department stores, Fleet Street indicates newspaper publishing, Lombard Street indicates banking, Leicester Square indicates theatres, and so on. The same characteristic occurs in other cities, of course; think of Bay Street in Toronto or Wall Street in New York.

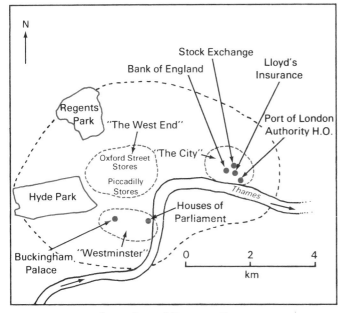

Figure 8-5. London's Central Business District

Inside the CBD there are several other aspects of city life in addition to the businesses just mentioned. Parks, for instance, provide very valuable open space in the otherwise heavily built-up area. For this reason they are known as the "lungs of London". Two of the largest are shown in Figure 8-5. There are also hotels for tourists (see Figure 8-6) who come to see old or historic buildings such as Westminster Abbey and Buckingham Palace. Tourists also come to see activities like the Changing of the Guard, or to visit theatres. There are central hospitals. There are cathedrals and universities, museums and art galleries, food markets, newspaper and book-publishing firms, and expensive apartments for rich people.

The concentration of business activity in central London is so great that the government has set up an agency called the Location of Offices Bureau (LOB), whose duty it is to try to persuade firms to move out of central London and into the suburbs. It has met with some success, but not enough. Usually what happens is that when a firm vacates central London, another one moves in to take its place! LOB is not winning. Why do you think business firms like to be close to the centre of things? (See if you can find out anything about *linkages*.)

The Changing of the Guard is a ceremony that takes place outside the Horse Guards barracks in Whitehall. What happens is that the Guards at the gates are changed: the new Guards ride up on horses, orders are barked, a ceremonial drill is performed, the tourists click their cameras, and the off-duty Guards then ride away. The new Guards sit still on their horses, the crowds drift away, and things calm down until the Guards are next changed.

Billingsgate is a fish market. Covent Garden is a fruit and vegetable market (and an opera house).
Smithfield is a meat market.

The Skyline Park Tower hotel.
Fine traditions in a beautiful setting.

Figure 8-6.   A London Hotel Advertisement. Tourists bring a lot of foreign money into London and are, therefore, an important source of "invisible earnings."

Marilyn Olsen

Marilyn Olsen

Buckingham Palace and the Horse Guards.

There is only so much land in the CBD, however, and there are many people competing to use it. Banks want to use it, insurance companies want to use it, stores want to use it, universities want to use it, art galleries want to use it, and so on. In fact, all the functions we have so far mentioned want to use it. Yet there is only so much land available. This means that only the very profitable firms or the very rich people or the very powerful government operations get a chance to use the land. Everyone else is squeezed out. Even the profitable firms, rich people, and powerful government bodies find that they must occupy cramped space, or build high-rises. London is an old town, however, so high-rises would look out of place. Furthermore, historic buildings would have to be removed to make room for them. Nevertheless, the need for more space is overwhelming. Rents are enormously high, existing offices are cramped, and it was even worthwhile for one developer to keep a high-rise office tower (called Centre Point) empty from 1965 to 1974 because the yearly rise in potential rents went up faster than he could have raised actual rents if the building had been occupied!

Just outside the Central Business District are the old residential and industrial areas of Zone 2. Some of these are very closely related to the Port of London Authority docks along the north bank of the River Thames. Food-processing and lumber industries are important examples of East End industries. The residential areas associated with the East End industries are generally old and often rather run-down. Some people would call them slums. But because of *urban renewal* schemes that have taken place in the last 25 years, some of the slums have been replaced by modern public housing. In the northern and western parts of Zone 2, however, housing conditions are different again. The houses were built to higher standards, and even though they are quite old they are still highly prized. Many well-paid professional people live in these northern and western parts of Zone 2, as for example in Kensington and Chelsea, which both lie just west of Hyde Park and Westminster (see Figure 8-5).

Zone 3 contains old residences, too, but in this zone the houses are often a little bit grander than the old houses of the East End. The Zone 3 houses were often built for what used to be called the "carriage trade" before the car was invented. They are usually quite spacious, with rooms for servants. Sometimes they have carriage drives (semicircular drives, so that a horse-drawn carriage could drive in and out without having to back up). Sometimes they haven't, of course, because even in the "carriage trade" days not everyone could afford a carriage with horses. However, the immense size of these old houses generally makes them difficult for most people to afford nowadays. As a result they are often split into separate flats or rooms for students, young married couples, and so on. Large parts of the boroughs of Hammersmith, Ealing, Lambeth, Southwark, Lewisham, and Greenwich are like this. It is chiefly into this zone that most of London's immigrants first come, because accommodation is usually cheap, and public transit serves the area fairly well. For example, Brixton in Lambeth Borough contains many immigrants from the West Indies; so do large parts of Camden Borough.

Zone 4 was built up mostly during this century, after cars and buses had become so popular as to permit the population to spread out more from the centre. In the land now covered by Zone 4 housing, costs were low at the

Marilyn Olsen

Some of the older houses in Zone 4 in North London.

Typical advertisements in a London evening newspaper (*The Evening Standard* or *The Evening News*) would read: SW2, best part, all convs., bed-sit, £10 p.w., 666 0150 after 6. Ealing, 3 rms, nr. tube, parking, £20 p.w., 555 2727 anytime. L'sham, 2 rms, b & k, no children, no pets, £18 p.w., 951 1212 before 5.

During the 1930s there was a huge building boom around what were then the edges of London. Not only was land cheap, but mortgage rates were very low. You could get mortgages then at about 3% or 4%. Most of the houses that were built were quite small, however.

time of construction (chiefly the 1920s and 1930s). For instance, a brand new 3-bedroom house, with a garage but without central heating, and a small area of garden around it would have cost the equivalent of about $2000 to $3000 in the 1920s and 1930s. This was mainly because the land itself, so far from the centre of town, was so cheap. In Zone 4, then, the houses usually have grassy backyards, and many (but not all!) of the streets have trees along the sidewalks. The streets themselves are not usually called streets, but *avenues, crescents, groves*, and so on, instead. The factories of this zone are usually well hidden behind screens of trees; inside, they are spacious and well-lit. They manufacture goods which became popular in the 1920s and 1930s, chiefly radios, car parts, domestic consumer items, and so on. Because this zone was relatively green and pleasant compared with the inner zones, population here grew rapidly, even after the central zones had begun to lose population (see Figure 8-7).

British Information Services

Private housing on the Courtlands Avenue estate in south-east London.

	1801	1821	1841	1861	1881	1901	1911	1921	1931	1941	1951	1961	1971
Inner London (zones, 1, 2 and 3)	1.0	1.5	2.0	2.7	3.8	4.6	4.5	4.4	4.3	4.0	3.4	3.2	2.8
Outer London (zone 4 and parts of 5)	0.3	0.3	0.4	0.6	1.0	1.9	2.7	2.9	3.7	4.6	4.9	4.8	4.7
Total (Greater London)	1.3	1.8	2.4	3.3	4.8	6.5	7.2	7.3	8.0	8.6	8.3	8.0	7.5

Figure 8-7.   Population Data for London, 1801-1971

Zone 5 is the present area of larger houses for the modern equivalent of the "carriage trade". Mixed in with the larger houses there are also plenty of standard homes, but throughout the area there is a general impression of green-ness. Trees and grass here contrast sharply with the old brown brick and blackened cement of the inner zones. Population is currently growing fastest in this outer zone, especially as Zone 4 people acquire more money and seek to move to a little better housing. In London, believe it, the housing with the most *status* is either a Zone 5 house or a Zone 1 apartment, and perhaps also some apartments and houses in certain parts of Zone 2. The in-between zones (two, three, four) are places that people like to move out of *if they can*. Figure 8-8 shows the pattern of population gain and loss throughout London, and you can see very clearly that it is the outer zone that is gaining population fastest. The innermost zone (Zone 1) is losing population, because even though most people would like to live there, only the very rich can afford to.

By now, you have probably decided that London fits

John Molyneux

People are still moving into the greener parts of Zone 4.

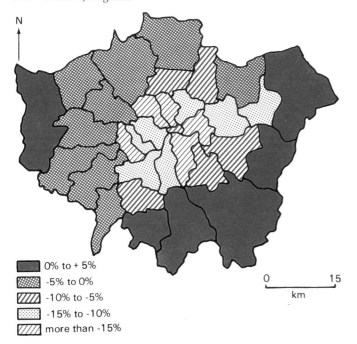

N

British Information Services

Privately owned detached properties in Haywards Heath, Sussex.

■	0% to + 5%
▨	-5% to 0%
▨	-10% to -5%
▨	-15% to -10%
▨	more than -15%

0 ———— 15
km

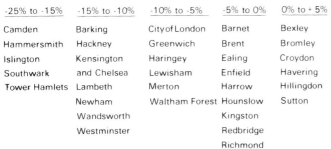

-25% to -15%	-15% to -10%	-10% to -5%	-5% to 0%	0% to + 5%
Camden	Barking	City of London	Barnet	Bexley
Hammersmith	Hackney	Greenwich	Brent	Bromley
Islington	Kensington	Haringey	Ealing	Croydon
Southwark	and Chelsea	Lewisham	Enfield	Havering
Tower Hamlets	Lambeth	Merton	Harrow	Hillingdon
	Newham	Waltham Forest	Hounslow	Sutton
	Wandsworth		Kingston	
	Westminster		Redbridge	
			Richmond	

Figure 8-8.   Percentage Population Changes 1961-71, by Boroughs

fairly well into the pattern proposed in the Concentric Theory. Nevertheless, you should be aware that there are still many differences we have not touched upon. After all, there are about 8 000 000 people in the London region, and the patterns they make on maps are not as simple as the Concentric Theory suggests. You must treat the theory merely as a handy way of examining towns, and not as a rule-book. To give you some idea of the sort of detail that cuts through the concentric zones of the theory, just look at Figure 8-9. This illustration is grossly over-simplified as well, because the real area of this section alone is approximately 65 km^2.

There are many postal districts in London, all of them prefixed by one or two letters. Thus, W1 is the first postal district in West London; it happens to co-incide with the expensive apartment, hotel, and department store area of the Central Business District, and therefore it is a highly prized postal address. Many people pay hundreds of pounds more a year just to obtain a W1 postal address. Other highly sought-after postal addresses include (somewhere in the new code numbers) W8, SW1, SW3, SW5, N6, NW3.

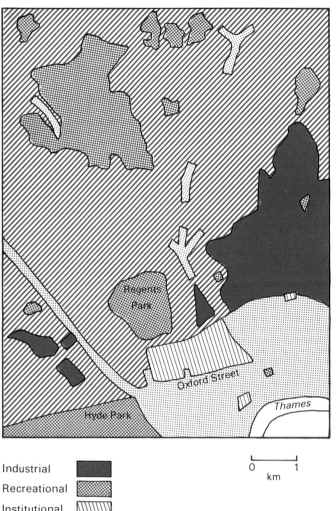

British Information Services

Trying to get to work in Oxford Street.

Industrial
Recreational
Institutional
Commercial
Residential

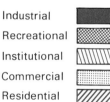

0   1
km

Figure 8-9.   Simplified Land Use in North-west London

QUESTIONS AND EXERCISES

1. How do high-rise buildings make more efficient use of CBD land than low-rise buildings?
2. What advantages does the CBD possess for offices and major stores as compared with the suburbs?
3. Draw a graph of the population data shown in Figure 8-7.
4. Why do you think the inner core of London is losing population?
5. What might the reasons be for the movement of population to the outer fringes of London?

9. MEDITERRANEAN LANDS

Christos Poulakis gave a grunt as he threw his weight behind the hoe. As it pushed through the dry earth, uprooting weeds and loosening the soil, he wondered if the beans would be as good as last year. They should be, he thought. He had used the same amount of fertilizer again, and the seed was of the same good quality. "All we need now is a little rain," he thought, slowly working his way up one row and down another. A shout made him look up. He should have known. It was his cousin, Nick. Nick had left his father's land in disgust last summer. He had grown tired of seeing tourists staring at him as he worked under a baking hot sun, peering through the viewfinder of a camera. He had gone to Athens and from there to Piraeus where he shipped aboard a boat that was bound for . . . where was it? Toronto.

He looked up at Nick's grinning face. "What are you doing back so soon? I thought you were there for good. Have you lost your job?" asked Chris.

"Oh no, nothing like that. I'm on holiday — came over on a plane and leave again after three weeks. I thought I'd rent a car and drive around Greece for a few days. Couldn't afford it when I lived here, and I don't know when I'll be back again. So why not!"

Chris scratched his head. "You certainly seem to be lucky. I've never had a holiday, and as for an airplane flight, well, it's out of the question. You must be rich now, all right. Wish I had left when I could have, a few years back when I was a bit younger. But Canada wouldn't take me. I had no education, no plumbing skills like you. No, I'm here for life, but my children are going to live in a city and hold down good jobs if I have anything to do with it. . . . Anyway Nick, come and see us before you leave. It isn't often anyone comes back once they've left the village."

As Lucio Pina tore around the corner on two wheels, he gave a whoop. The Fiat was a good car for that sort of thing, even in the streets of Turin. He had been able to save some money working regularly, something he had been unable to do in Catanzaro. The muffler popple-pop-

popped as he tore across an intersection, dodging motor-scooters and people. He grinned as he remembered Umberto's joke: "There are only two kinds of pedestrian in Turin. The quick and the dead!" He would never return to Catanzaro to live. The South was not for him, this he had always known. But maybe he would holiday down there. Everyone would be so jealous of his good fortune!

Fernando Costa had a look on his face as sour as the vinegary wine he had had with his meal. Portugal was allowing the Empire to break up! How was he to break out of this fishing village now? He had accepted a tour in the army, and now Angola and Mozambique had been given their independence. They might have waited for a couple of years, at least until he had seen them. Why, he might even have decided to settle there. At least it would have been a change. One way or another he was determined to leave the village in which he had been born, 19 years before. All his father did was talk about the poor fish catches. Well, if that was true, why didn't he leave and find a job in Lisbon?

At the thought of Lisbon, his eyes lit up. No more tending the goats. No more hard work for little return. He would just announce that he was going to Lisbon to find work, see a bit of life, meet a few people. It was an exciting prospect. All he had to do was go. Lots of young men his age had already left the coastal village. They were not content to work in the large hotels that had sprung up recently. And neither was he! His mind was made up. Tomorrow. . . .

If the 3 men you have just read about were to meet, they could not understand one another. They speak different languages. But in many other respects, they are very similar. It is all these common factors that we are going to explore when we investigate Mediterranean lands. The area under consideration is shown in Figure 9-1.

From the earliest times, the Mediterranean Sea has allowed civilization to progress. It was at once a routeway, probably the most important one, between nations; it was a defence, a source of food, even at one time a god. The land areas that came under the domination of this sea were all peninsulas or islands. Mountainous and rugged for the most part, these land areas could communicate best with others by sea. Because navigation in those far-off days was so chancy, voyages were usually made

Cars and Rome do not mix very well. Traffic vibrations have damaged the Colosseum, the noise of horns is deafening, the rush-hour traffic jams exasperating. You can understand, then, why Rome has now banned cars in certain parts of the central city. Julius Caesar would have laughed. He tried to ban chariots from the city in 51 B.C. Did he succeed?

The humble sardine once swam in large shoals near Portugal's shores. As time passed, the shoals moved farther out into the Atlantic, and now the traditional *faluca*, a simple wooden sailboat, cannot reach them easily. Expensive motor trawlers are required, but the fishermen are poor.

The rocky coast of southern Greece has many ancient wrecks. One that was discovered about a century ago was a Roman ship carrying marble and granite *sarcophagi*. These were large coffins designed to pay tribute to the dead of old Rome. They were ancient status symbols. Other wrecks have been found carrying wine in *amphorae*, large two-handled jars. The wine did not taste good, however. Even though the sea did not get in, the wine decomposed during its 2000-year storage.

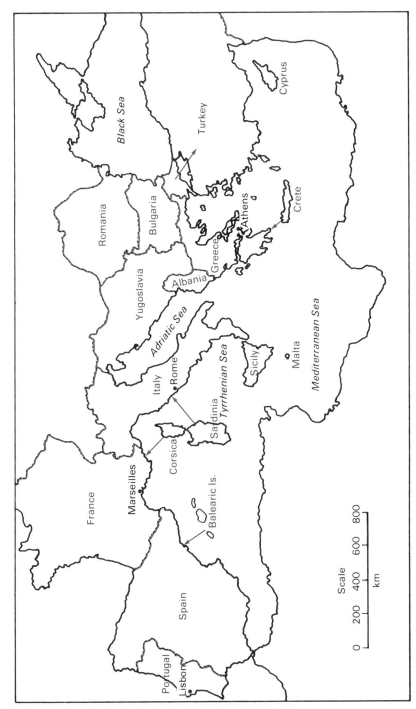

Figure 9-1. Countries and Places Mentioned in This Chapter on the Mediterranean

within the sight of land, just where the water was shallowest and most dangerous. When a storm sprang up, disaster was often inevitable. Ships, cargoes, and crews perished. Today their remains form an important source of information about life around the Mediterranean for thousands of years. Copper ingots from Syria, marbles and bronzes from Greece en route to Rome to grace a senator's garden — all these things show the ancient desire of people to civilize and enrich, but also to exploit and change.

At one time, the Mediterranean lands were rich and fertile, cloaked in forests of ancient oaks and beeches, elms and cypress, holly and sycamore. But early settlers and later conquerors misused the land, and it gradually decreased in productivity. Goats and sheep were introduced to graze on the hillsides, and they destroyed growing plants. The soil that was now exposed was quickly washed to sea, being no longer kept moist and held together by plant roots. Great, gleaming slopes of limestone appeared, unable to hold water or grow very much. The forests shrank. The numbers of people grew. Commerce increased. The region prospered.

THE TRADITIONAL RESPONSE

In most parts of the world there is a possibility for people to make a living. How they do so varies a good deal. For one thing, climates are very different. How high are the temperatures, and how much rain falls? Is there a good distribution of rainfall throughout the year or is it seasonal? What is the soil like? Which crops grow best and produce most? In other words, is the climate favourable or not? Are there any raw materials in the form of natural resources like coal, oil, iron ore or lumber? Are there any links with the rest of the world?

In talking about the Mediterranean region, the best way to answer questions like these is to see what actually grows naturally and what type of farming is carried on. After all, people have lived here for many centuries and have learned through experience what grows best. The land and the natural conditions are fixed and pretty well unchanging. How have people responded to these conditions, and even changed them?

As you can see, the Mediterranean has a distinct type of climate. The winters are cool and wet, summers hot and dry. Perhaps it would be more accurate to say that only

	J	F	M	A	M	J	J	A	S	O	N	D	
Marseilles	7	8	11	13	16	20	22	22	19	15	11	8	temperature (°C)
	42	36	48	56	43	28	18	20	61	97	71	53	precipitation (mm)
Lisbon	11	11	12	15	16	19	22	22	20	17	14	11	temperature (°C)
	91	89	86	66	51	20	05	05	36	84	109	104	precipitation (mm)
Athens	8	8	11	15	19	23	27	27	23	19	14	11	temperature (°C)
	51	43	30	23	20	18	08	13	15	41	66	66	precipitation (mm)
Rome	7	8	11	14	18	22	24	24	22	17	12	8	temperature (°C)
	81	69	74	66	56	41	18	25	64	127	112	99	precipitation (mm)

Figure 9-2.   Mediterranean Climatic Data

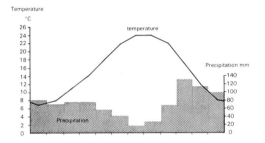

a little rain falls in summer, and that evaporation is very great under the blazing hot sun.

The type of plants grown must be able to resist extremes of temperature and survive drought that might last for months. Larger plants will need a thick bark that will act as insulation. The root system will be widespread and go deep into the cooler, moister earth. In summer when the plants are growing, and are therefore covered with leaves, water loss through the leaves by a natural process called *transpiration* is at a maximum. Therefore the leaves should be small in surface area, preferably covered with a waxy coating that still further cuts down on water loss. Traditionally there has been reliance on two such plants.

The olive tree, *olea europea*, is often taken as an indicator of a Mediterranean climate. Its bark is thick, and the narrow, shiny leaves allow it to survive summer drought easily. Olive trees are grown in orchards known as *groves*. The young, immature olives are green and a little sour to the taste. They are treated with alkali to get rid of this tartness, then stored in brine. Black olives are ripe. They are mostly crushed for the oil they contain. Olive oil forms an important source of oil and fat in the Mediterranean lands, and is used widely in cooking.

The cork oak is another *xerophyte* — a plant adapted to survive dry conditions. It produces in its bark a layer of tightly fitting cells that die, accumulating a little more each year. This layer is cork — the material used for stopping bottles. Every 9 or 10 years, the cork is stripped from the tree — carefully, in order not to injure the sensitive live cells beneath it. Although the cork oak can make do with very little rainfall, it must be stripped when humidity is high. For this reason, Portugal, facing onto the Atlantic Ocean and intercepting damp air from the winds that blow off it, has the greatest output of cork.

Marilyn Olsen

A Spanish olive grove.

Cork provides a living for more than 20 000 workers in Portugal. Over 600 factories process the material into wine-bottle corks, life preservers, and a wide variety of other things, including decorative wall panels. The use of foam plastic has seriously hurt the cork industry, though, and it has had to explore the possibility of new uses for cork.

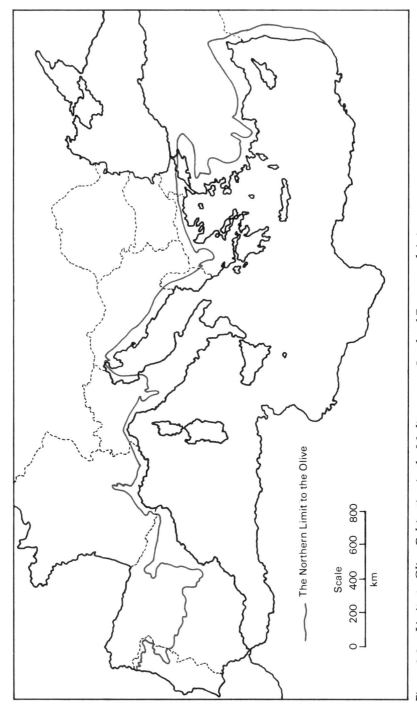

Figure 9-3. Limits to Olive Cultivation in the Mediterranean Lands of Europe and Asia

The Northern Limit to the Olive

Scale

0   200   400   600   800

km

Perhaps the most familiar plant of the Mediterranean lands is the grape vine, *vitis vinifera*. This plant originated in the Caucasus Mountains of the U.S.S.R. and was imported to the Mediterranean countries thousands of years ago by migrating peoples. When North America was colonized, settlers took the familiar plant with them and planted it there. It thrived, only to be nearly wiped out by the pest *phylloxera*, which attacked the roots. *Phylloxera* were also carried to Europe and nearly wiped out the wine industry there, too. It was found that hardy rootstocks of native North American vines could resist the pest. Growers grafted *v. vinifera* onto these rootstocks, and in this way, the European wine industry was saved.

Most of the wine produced is rough and of not very good quality. It is largely drunk within the home countries. Generally, only the better quality wines are exported. Good *vintages* of wine are the result of particular climatic and weather conditions, as well as soil type. Because climate and weather vary from year to year, it is possible to speak of "good years" or "bad years" for wine.

Besides being used for wine, of course, grapes can be sun-dried to make raisins and sultanas. Crops of other fruits can be produced, as well, including grapefruit, oranges, even bananas. With irrigation, crops of rice are common in areas like the Po River valley of northern Italy, the Rhone delta of France, and various valleys of the Iberian peninsula. You may think that rice is rather an exotic crop, but, in fact, the Mediterranean countries are exceedingly good at producing it. In a good year, France can achieve yields in excess of 4000 kg/ha, Italy can top 5000 kg/ha, and Spain can produce over 6000 kg/ha. By way of contrast, Japan's best rice yield was 5847 kg/ha, India's 1717 kg/ha, China's 3137 kg/ha, and Bangladesh's 1776 kg/ha.

A critical weakness of Mediterranean agriculture, at least the traditional sort, is the small size of agricultural holdings. In Italy it is only 7 ha per farm, in Spain 15.6, in Greece 3.2, and in Malta 1.5. Yet it has been calculated that a modern European crop farm needs to be between 81 and 121 ha in size for it to operate commercially and make a profit. Now while in Italy, Greece, and Spain there are obviously some farms that *are* of this size or maybe even larger, most of them are much smaller than this. So they are not very profitable. On a small area it is often impossible to use machinery, even if the farmer could

1972 was a poor year for wine. Spring was late and cold; August was wet, with little sun, and the grapes rotted instead of maturing. It was a disaster for Italy, where each person drinks an average of 100 bottles of wine a year. About 300 000 families depend on wine for their income, while another 1 500 000 people are involved in processing and marketing it.

One of Mussolini's achievements was to drain the Pontine marshes outside Rome and turn them into fine farmlands. In doing so, he also got rid of the *Anopheles* mosquitoes which spread the disease called malaria.

	Olive oil tonnes	Wine thousand tonnes	Raisins tonnes	Grapefruit thousand tonnes	Figs thousand tonnes	Oranges thousand tonnes	Lemons & limes thousand tonnes	Tomatoes thousand tonnes
Albania	5 000	10			13	4	1 050	
Bulgaria		345						730
France	1 500	5 854			4	4	1	490
Greece	255 000	489	140		150	400	140	1 100
Italy	400 000	5 919	0.6	2	177	1 490	726	3 262
Malta		2						6
Portugal	51 605	726	0.1	7	200	121	11	910
Romania	no data	600						900
Spain	485 000	2 645	5	6	120	2 493	165	2 017
Yugoslavia	1 500	626			22	2		372
Cyprus	2 000	49	5	55	2	130	450	24
Turkey	150 000	52	315	10	220	528	4	2 000
World total	1 551 280	27 815	806	3 552	1 244	27 512	3 536	32 190

Figure 9-4. Production Figures for Some Characteristic Crops of Countries in the Mediterranean Region, 1972

afford it. There is certainly not room enough on such farms to produce large cash crops. Much of Mediterranean farming is subsistence, or part subsistence farming; a farmer eats mainly what he grows, and relies for his cash income on what he sells at the local market. Why does this state of affairs prevail here, when other countries on the same continent, such as Britain, have average farm sizes of 40 ha or more?

Part of the answer must lie in the physical geography of the region. Much of the landscape (and here you really must open your atlas) is mountainous. Communication is thus difficult, especially since the countries of Spain, Greece, and Italy are peninsulas. Their southern portions especially tend to be remote and isolated. Just look at the positions of Sicily, Sardinia, and Corsica, the islands in the Aegean Sea, and places like Calabria in Italy, or the Peloponnesus in Greece. They are not easy to reach, tend to have low standards of living, and are subject to emigration as a result of overpopulation.

Shell Photographic Service

An Italian Alpine village.

THE MODERN RESPONSE

For many years Italy has worried about the development of the south. In fact, ever since unification of the country in 1871, the south has lagged behind the north in industry and investment. For example, over a century ago, the Kingdom of the Two Sicilies had a total of 98 km of railroads, while Piedmont in the north had 803 km. Most of the imports of the Two Sicilies consisted of finished goods,

The *Mafia* started life hundreds of years ago as a secret organization to prevent the exploitation of the peasants of Sicily and southern Italy.

but in Piedmont, over 30% of imports were raw materials and machinery. In the south, death rates were high, as was infant mortality. Even so, enough people survived into adulthood to swell the ranks of the work force so that unemployment was high. Malaria, spread by mosquitoes on the swampy coastal plains, was bad enough to drive most people into the mountains. There the poor soils and shortage of land allowed mere subsistence at the best of times.

The situation did not improve over time. In fact, it grew worse. Between 1914 and 1939, the output of southern agriculture increased 3%. In the north, it rose 22%. *Literacy* rates in the north climbed from 32% to 70% between 1861 and 1911. The corresponding figures for the south were 13% rising to 40%. What held the south back so much was the poor agriculture practised there. Unlike farming in the great valley of the Po in the north, modern practices of farm ownership and production did not take place. Most land was owned by absentee landlords, who were not interested in anything but the incomes they were receiving.

During 1896 to 1936, Italy achieved economic "take-off." The country was able to finance and expand industrially, using its own capabilities. However, while industrial income (industry was mainly concentrated in the north) rose by two-thirds, agricultural incomes actually fell. In the south there were no large industrial cities to absorb a flood of peasant farmers if they wanted to leave the land.

By 1948, unemployment in the south stood at 16.6%. In the north the figure was 7%. One million people were without work in the south. Millions more were underemployed, trying desperately to wring a living from "farms" that were too small, infertile, and improperly used. The government responded by setting up the Fund for Extraordinary Works of Public Interest in Southern Italy. In Italian, it was known as the Cassa per il Mezzogiorno, or more simply, as *Cassa*. Basically, it was to provide water control and irrigation, improve the land, build aqueducts and sewers, encourage the mechanization of agriculture, build factories to process agricultural products, and encourage tourism. What it meant to the people of the south was hope for the future. No longer would their youth be obliged to go to the factories of the north or to the New World.

By the late 1960s one big weakness of the Cassa's plans

Shell Photographic Service

A castle in San Marino, legally a separate state of that name in Italy.

	1950 to 1960	1952 to 1964	1957 to 1962	1959 to 1974
Agriculture	77%	69.4%	55.8%	55.3%
Transportation, Communication	9%	14.8%	12.7%	12.6%
Aqueducts, Sewers	11%	13.8%	15.3%	15.0%
Industry	—	—	12.0%	11.8%
Crafts	—	—	0.2%	0.2%
Tourism and Hotels	3%	2%	2.2%	2.7%
Fishing	—	—	0.2%	0.2%
Education and Technical training	—	—	1.5%	2.1%
Social Welfare Agencies	—	—	0.1%	0.1%
Totals	1000 billion Lire	1280 billion Lire	2053 billion Lire	2078 billion Lire

Figure 9-5. Investment Plans for the Cassa per il Mezzogiorno

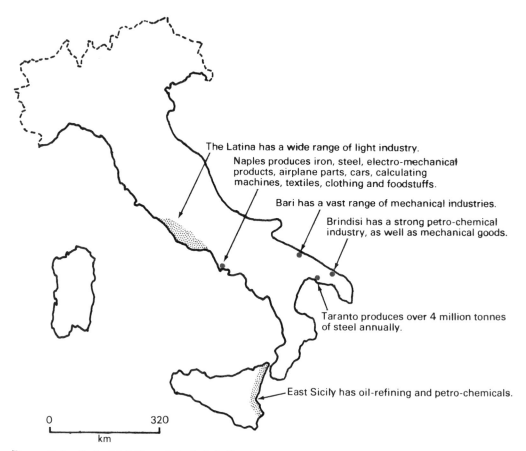

The Latina has a wide range of light industry.

Naples produces iron, steel, electro-mechanical products, airplane parts, cars, calculating machines, textiles, clothing and foodstuffs.

Bari has a vast range of mechanical industries.

Brindisi has a strong petro-chemical industry, as well as mechanical goods.

Taranto produces over 4 million tonnes of steel annually.

East Sicily has oil-refining and petro-chemicals.

0   320
km

Figure 9-6. Industrial Growth in Italy's South

was apparent. It was not enough simply to provide factories, roads, a trained work force, and all the other similar things known as *economic infrastructure*. Industry itself had to be persuaded to set up operations in the south. If it would not go voluntarily, then it must be directed.

The prime target for such direction turned out to be the large state corporations, such as ENI (Ente Nazionale Idrocarburi, the oil company). More than 45% of their new investment in the 1960s took place in the south. Private companies only invested 22% of their new money in the south. By 1965, the Institute for Industrial Reconstruction, ENI, and the National Agency for Electric Energy were required to spend 60% of their investment in the south. All state corporations, they control most of the country's production of steel, motors, oil, aluminum, as well as shipping, railroads, and electrical production. They produce about one-third of Italy's Gross National Product. The impact they have had on Southern Italy can be seen in Figure 9-6.

By 1970, $11 billion had been spent. Fiat, Pirelli, and Olivetti all had plants in the south that sent exports all over the world. However, in order to reduce emigration (nearly a million people left between 1961 and 1967), the Italian government is now spending money to encourage *foreign* companies to set up operations in the south, too. Government relaxation of tax laws is designed to cut wage bills by 50%, and wages make up over 70% of costs in modern industry. In addition, in 1972 and later years, the large state corporations had to invest 80% of their new capital in the south.

Italy is not the only country in the Mediterranean region to try to solve its economic problems. Most of the other countries are also trying, under a variety of governments, ranging from a right-wing dictatorship in Spain to the communist regime of Albania. One and all have tried tourism as a means of instant rises in income. They depend heavily upon people from Western Europe and North America. Ancient and modern history, ruined castles and inhabited castles, tremendous variety in food, culture and landscapes, a benign climate — the Mediterranean has it all. Islands, mountains, lakes, rivers, beaches, something for everyone.

But tourism is not the complete answer on its own. Tourism merely provides foreign exchange that must be used for investment in developing the economies of countries such as Greece, Yugoslavia, and Portugal. It is not

Marion Jones

The Parthenon (above) and the temple or shrine of the Caryatids (below) are great tourist magnets on the Acropolis, Athens.

Bonnie Eccles

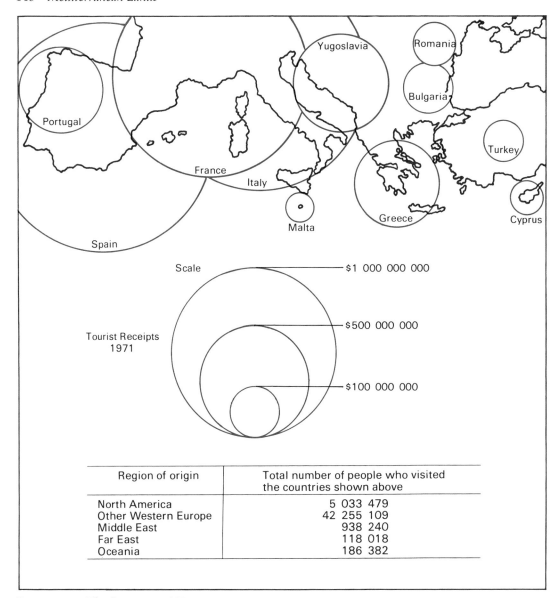

Figure 9-7. The Structure of Tourism, 1972

Region of origin	Total number of people who visited the countries shown above
North America	5 033 479
Other Western Europe	42 255 109
Middle East	938 240
Far East	118 018
Oceania	186 382

easy, for traditionally many able and skilled people have felt that, to be successful, they had to emigrate to countries such as Canada and Australia. Countries with small populations such as Portugal, with only 10 000 000, feel seriously threatened by emigration. So does Greece, with only 9 000 000 people.

One country that has seen great economic growth is France. The Mediterranean coast of France is traditionally a playground for tourists. Today the advent of the touring

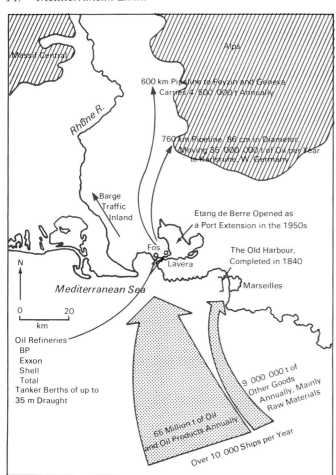

Figure 9-8.   **The Position of Marseilles, France**

Portugal "lost" over 100 000 emigrants in 1971. They went illegally, said the government. Youths and their families determined to avoid the draft just took off, 6983 of them to Canada. Only 10 made the return journey. Portugal wants its skilled and educated people back. The country is on the edge of the Common Market and needs these people in order to build up industry under the much-vaunted 6-year development program, which ended in 1974. Portugal exports mainly wine, cork, sardines, textiles, and minerals.

motorist, living in a tent or perhaps a cheap *pension*, has resulted in a tremendous increase in the number of visitors who come to see the countryside. Yet side by side with the tremendous attraction of the French Riviera is the port of Marseilles.

Marseilles began life thousands of years ago as a colony of Greece. Of course, it changed hands many times and grew with the ages. It is admirably suited to command the inland route to France and the best route to northern Europe. The largest port in the Mediterranean, it did a total trade in 1970 of 74 million tonnes of goods. That surpassed Genoa in Italy, which had only 52 million tonnes of goods passing through in 1970, and every year the gap grows wider. Genoa has to communicate with the rest of Europe across the Alps. Marseilles is near the mouth of the

Rhone River, the only major European river emptying into the Mediterranean. By 1976, 3000-t barges will be able to reach Lyons, in the centre of France, 300 km away. Electrified railroads speed cargoes inland. A high-speed freeway system permits a fast trip to Paris. Eventually, Marseilles will be linked directly to the German *autobahn* system and the Italy-Spain freeway. In addition, there is a major international airport only 19 km away.

Because there is little tidal range in the Mediterranean, the port's use is simplified. It has grown steadily as an important trade centre since the Middle Ages and today takes in the little town of Fos, 50 km to the west. There are 35 km of docks built and planned to accommodate tankers of up to 1 000 000 t. The berths are ready, but the tankers have not yet been built!

An unpleasant consequence of so much industrial activity around the shores of the Mediterranean Sea is *pollution*. The sewage systems of places like Cannes on the French Riviera simply cannot cope when the summer population increases from 60 000 to 300 000. At Hyères in 1971 the police actually closed the beach because of the hazard to swimmers in the form of coliform bacteria.

Cities have grown greatly. Athens is typical of many cities that see many tourists each year. In addition, it has a nearby port, Piraeus. The Greek merchant shipping fleet stands third in the world in the number of ships. Piraeus and Athens have grown a lot in recent years as international trade has increased, for the Greeks act as freight carriers to the world. Athens has now reached the point where it will have to do something to solve its traffic congestion. The city has increased its population by six times this century and could well grow a lot more as Greece searches for oil in the Aegean Sea. It already has made one find near the island of Thassos, and Athens-Piraeus is ideally situated to act as a supply centre for this industry.

The Mediterranean region has two faces, then. One is the traditional way of life, especially in the rural areas in the south, which can support only limited numbers of people. Tourism was a short-term answer to solving the problems of these economically depressed regions, but the only permanent solution seems to lie in the new face that the Mediterranean is presenting to the world. It is a bustling one, lying astride the ocean route from the Suez Canal, which was reopened in 1975. With this, the Mediterranean will become an oil artery for western

In 1973, the famous oyster beds of Naples were closed to shell-fishermen. The reason? Inadequate sewage treatment had rendered the shellfish a threat to human health. Naples had more cases of cholera in 1973 than it had had for a long time.

One estimate says that the Mediterranean will die before the end of the century, its life-support potential choked out by the sewage, pesticides, fertilizers, and detergents that enter it every day — and, oh yes, do not forget the oil slicks.

Greece is running short of seamen. The country's ships need 135 000 sailors. There are only 115 000. In 1962, 1519 Greeks took out their seaman's papers. In 1972 there were just 30. The gap is bridged by importing foreign sailors.

What does Athens want, an overhead commuter train to ease the rush-hour congestion, or a subway? It looks as though the former suggestion will win. Although the elevated train service might look a little out of place, it is only half the cost of a subway, and a subway might do more damage to the city's historic sites than is acceptable.

Europe, even more so than today. With it will come more factories and industrial complexes, which represent prosperity. It remains to be seen whether the old and the new can go hand in hand. Let us hope so.

Italian cooking often uses two ingredients that are found everywhere in Italy: tomatoes and onions. Here is a popular Italian dish:

Spaghetti Bolognese

Sauce:	2 ml oregano
30 ml olive oil	1 ml marjoram
170 ml chopped onion	1 ml thyme
1 clove garlic, crushed	1 ml crushed rosemary
1 large can tomatoes	2 ml salt
1 can tomato paste	0.5 ml pepper
1 bay leaf	7 ml sugar
2 ml basil	226 g ground beef

Cook onions and garlic in olive oil until soft. Do not let them brown. Add remaining ingredients except meat, mixing well and breaking large pieces of tomato. Simmer uncovered for 1½ to 2 hours, until thick. There should be about 680 ml of sauce.

Now brown the ground beef in a separate skillet. Pour off fat. Add meat to sauce. Cook for 20 to 30 minutes longer.

Cook spaghetti according to package directions. Drain well. Top with sauce, Parmesan cheese, and hot peppers if desired. Warm, crusty rolls are good with this dish. Makes 4 to 6 servings and freezes well.

Spain feels that it has too many tourists. Thirty-six million foreigners visited the land in 1974, and Spain's own population numbers only 34 million. As the standard of living rises in Spain, Spaniards themselves who want to take a trip find it difficult to find hotel rooms during the summer. They also complain that the Mediterranean coastline has been ruined in Spain by the mushrooming of high-rise hotels.

QUESTIONS AND EXERCISES

1. Find out what you can about Greek colonization of the Mediterranean thousands of years ago.
2. Complete the climate graphs for Lisbon, Athens, and Marseilles. The information for this is in Figure 9-2.
3. Devise a way of showing the information in Figure 9-5.
4. For which crops in the list for Figure 9-4 is the Mediterranean region most suited? Construct a divided circle to show the world production of each crop.
5. Cook *spaghetti bolognese*, above, for your family and tell them why it is a characteristic Italian dish.
6. On an outline map of the Mediterranean, construct proportional-width arrows to show the numbers of tourists from the regions of origin in Figure 9-7.

10. OVER POPULATION, POVERTY, AND HUNGER

SUPPORT OF LIFE

Indian Government Tourist Office, Toronto

Rajasthani cowherd in N.W. India. Does this represent overpopulation?

No one knows exactly what overpopulation is. It is certainly not just a lot of people crowded into an area. If that were all it was, then all the world's cities would have to be counted as overpopulated, and that is not true. So would certain crowded countries, such as the Netherlands, and that is not true either. Overpopulation means more than simply a large number of people in an area.

It is not easy to say what the other factors are that make up overpopulation. Almost certainly, poverty and hunger are two of the most important. If the people in an area work very hard for a living and still go hungry, then we can probably say that area is overpopulated. In other words, the people (for some reason or other) are not capable of using their resources well enough to yield an acceptable standard of living. Even this statement is not very exact, because we have to ask, *acceptable to whom?*, and we shall not get an answer that is agreeable to everyone. Certainly, even in any one country, people will all have different ideas of what is acceptable. Nevertheless, for the time being we will define overpopulation as an imbalance between population and resources that is characterized by hunger and poverty.

HUNGER

You all know that at least half the world's population is permanently hungry, and that from 25% to 30% are actually starving. Even many little children who cannot read or write know this. For all they know, everyone in the world may be starving, for starvation is all they ever see around them — day after day, week after week, month after month, year after year, generation after generation.

It is fairly easy to use an approximate measure for hunger. The United Nations publishes figures every year showing the number of Calories that are available to each person in each country in the world. The higher the figure, the more food there is available; the lower the figure, the less food there is available. Countries with fairly high figures, say over 3000 Calories per person per day,

Ask your school librarian to order copies of the UNESCO *Courier*. They are available through the nearest Information Canada office. If your school already receives copies, then read some of the articles on population.

A generation is the rather vague length of time that is determined by how long it takes for a *cohort* of babies to grow to an age where they can themselves marry and have children. Thus a generation is approximately 20 to 25 years.

A *cohort* is the official name for any large group of people whose birthdates all fall within a year or two. Thus all the people born in 1961-62 form a *cohort*. So do all the people born in 1962-63.

How many Calories do *you* eat a day?

can be considered to be well off and not overpopulated. On the other hand, countries with low Calorie figures, say under 2500 Calories per person per day, are likely to be considered hungry, and thus overpopulated. Figure 10-1 shows the situation across Eurasia. Southern Asia, right across from the Middle East to China, stands out as a hungry part of the world, even allowing for the fact that people do not need to eat quite so much in hot areas as they do in colder lands. Europe and northern Asia, by comparison, show up as well-fed regions.

One very important fact emerges from Figure 10-1: Hunger has very little to do with high population density. Some countries with a high population density also have a high Calorie figure (e.g., Netherlands, Belgium, U.K., Germany), while others with a high population density have a low Calorie figure (e.g., Pakistan, Bangladesh, India). Furthermore, some countries with a very low population density figure still have a high Calorie figure (e.g., U.S.S.R., Finland), while others with a low population density also have a low Calorie figure (e.g., Khmer, Malaysia, Laos, Burma). The cause of hunger is clearly not a high population density. That's worth repeating; crowded lands do not necessarily support only starving people, and uncrowded lands do not necessarily support only well-fed people. This is probably a new idea to you. After all, you have probably been brought up to believe that the world's population growth was likely to be one of the major causes of starvation. So think about it for a moment, and get used to the new idea.

You can measure the efficiency of farming by either food output per hectare or food output per farm worker. By both of these tests, much of southern Asia has inefficient farming, and hunger is much more the product of inefficient farming than it is of a high population density. There are many possible reasons for inefficient farming, such as resistance to change for religious reasons, inability to experiment with new methods because of debt or poverty, lack of knowledge of new techniques, scarcity of new seeds, fear of *anything* new, persistence of traditional systems of landholding, security of old ways, and so on and so on. Before farming can become efficient — and so produce much more food per hectare or more food per farm worker — all these problems have to be overcome. As you can imagine, that is an enormous task, and it progresses only slowly. In the meantime, hunger continues. Indeed the problem gets worse, because the coun-

Approximate population densities (people per square kilometre):

Netherlands	400
Belgium	302
U.K.	236
West Germany	241
Pakistan	75
Bangladesh	550
India	168
U.S.S.R.	11
Finland	15
Khmer	38
Malaysia	33
Laos	13
Burma	42

Rapid population growth is not necessarily taking place in countries with a high population density. In some it is; in others it isn't. Nor is a high population density a necessary condition of rapid population growth. Rapid population growth occurs in *some* countries with a high population density, but it also occurs in others with only a low population density. In other words, a high population density and rapid population growth have no necessary relationship. But whatever the population density, rapid population growth will make it denser.

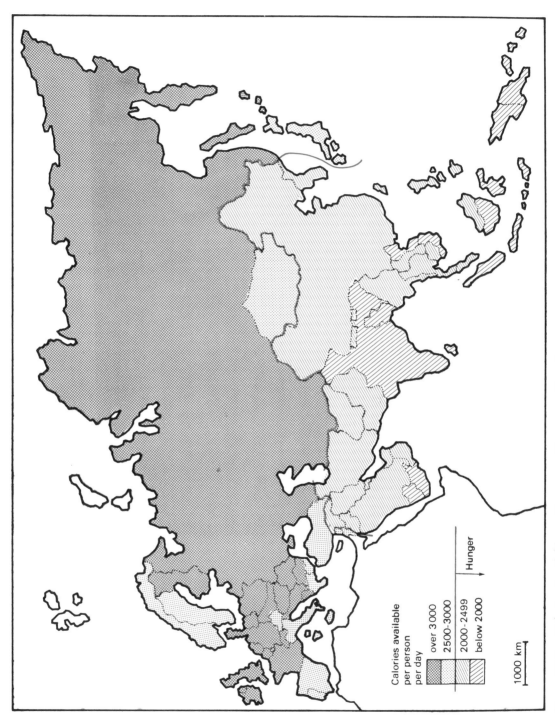

Figure 10-1. Calories Available per Person per Day across Eurasia

tries already suffering from hunger, because of inefficient farming, are also the countries that are having very rapid increases in population. This is the basis of what you previously learned about hunger, that the population explosion will cause hunger. It is not so, of course; the hunger would exist anyway. The population explosion just makes solving the problem much more difficult — and it does produce a lot more starving people. However, even if there were no additional people in southern Asia, or even if the numbers were halved, there would still be hunger. The problem is not caused by numbers, but by inefficiency, and throughout southern Asia farming is inefficient.

It is not an easy problem to solve. Indeed, it may even be the worst problem the world has ever faced. As you are undoubtedly aware, there are many views about what should be done. Here are some of them.

Usually a lot of changes are needed to make inefficient farming efficient. But before there can even be any changes, certain factors must be present. There must be the *desire* of the farmers to change, and there must be *help* and *money* available, usually in the form of *foreign aid* from another country. Once these basic factors are present, then the hungry farmers can get on with changes themselves. One of the most urgent changes they must usually make is in their use of fertilizer. There are three elements that are absolutely essential for all plants: nitrogen, phosphate, and potash. Years of inefficient farming drain the soil of these elements, and crops become poorer and poorer. Fertilizer restores them to the soil. Just look

Shell Photographic Service

Arabian fishermen in Muscat and Oman. Does this represent overpopulation?

Tractors are not tools; they are power units. They replace animal and human labour. They do not actually plough fields or dig ditches or harvest crops; they provide the power to do so.

It seems silly to question the *desire* of starving farmers to change their ways. Not so. Obviously the farmers want to change from starving to not starving, but they are not always keen to change their *farming methods* to meet this desire. You can desire things all you want, of course, but you will not achieve your goals unless you put deliberate effort into your work. For many starving farmers, change is a big gamble, and it is the farmer's family's lives that are at stake. If the new ways do not work, then the farmer wonders what will become of his family. At least with the old ways he knows that there will be some food, even if not enough. But with the new ways, who knows? Just because a stranger from a foreign land says so? Perhaps it is wiser not to risk it! Better to go hungry but remain alive than risk failure and starve to death.

	Input		Output
	Potash fertilizer kg per person per year	Tractors per million population	Calories per year per person
China	2	220	2 050
India	4	140	1 940
Indonesia	2	85	1 750
Denmark	363	35 000	3 140
France	278	24 800	3 270
West Germany	197	22 850	3 000

Figure 10-2.   Selected Agricultural Input-Output Data for Six Countries, 1971

at the potash data in Figure 10-2. The very heavy use of this fertilizer in well-fed European countries stands in stark contrast to almost no use at all in hungry Asian countries. It is just the same with the other elements.

Another very important change that must be made is in the increasing use of farm machinery. Tractors are the most basic type of farm machinery, and again Figure 10-2 shows the tremendous difference between European and Asian countries. The hungry farmers must gradually begin using more machinery and fewer people to work the farms. (The people who were not needed on the more efficient farms could then go to the towns to work in fertilizer and tractor factories. Of course, people in the towns should set up these factories so that everything goes smoothly. Otherwise there will be massive unemployment, and perhaps riots — and no fertilizer or tractors.)

As well as these changes in the use of fertilizer and machinery, hungry Asian farmers also need to start using new seeds and new techniques. New seeds are available

When the first experiments were carried out in breeding miracle rice at the International Rice Research Institute in the Philippines, the chief idea was to produce rice that had larger ears. Unfortunately, the experiments were so successful that the rice plants became top-heavy and all fell over. Harvesting was impossible. Then the institute had to devise ways of breeding rice that not only had larger ears, but would also stand up straight under its own weight. Thicker stems were not the answer, because this took too much nutrient from the ears. Eventually the researchers came up with rice plants that were shorter and stubbier, but still with large ears. Miracle rice was born.

Sue Mason

There are many problems to be overcome.

Sue Mason

Is this the problem?

Mick Horner

Does this person have the desire to change his ways?

Mick Horner

An inefficient one-man irrigation system in Iran. Does this represent overpopulation?

(miracle rice, for example), but they are expensive, and the poor farmers cannot afford to buy them unless they get foreign aid. The new techniques that must be employed include such things as regular pumped irrigation (instead of gravity irrigation), water storage (for provision of water during dry seasons), pesticides, crop rotation schemes, and so on. There must also be vastly improved crop storage facilities. Did you know that when all the crops have been harvested, insects eat one-third before the hungry people get a chance to?

There must also be some successful attempts at land reform. Most hungry farmers in south Asia do not own their farms; they are tenants, paying rents to landlords. Often the rents are so high that as much as half of a farmer's crop is used to pay rent. Thus, when harvests are poor (as they often are on inefficient farms) many farmers are tempted to go to the moneylender. Debt, therefore, is another major problem, for who can afford to experiment with new techniques when he is loaded down with debt? Another problem with the land is that frequently the actual holdings are too small for efficient use. It is usually not worth buying machinery and setting up costly pumped irrigation systems when the farms are small. Several small farms need to be *amalgamated* to be worked efficiently. Here too the farmers whose land has been bought out must go to the towns to seek work in fertilizer and tractor factories (which are often not yet in existence).

Until these problems of inefficient farming are solved, there will always be hunger in south Asia, regardless of how many people live there.

POVERTY

Poverty expresses itself in many, many ways. It is not being able to pay for shoes, to take vacations, to afford entertainment, to buy furniture, to rent a home — in fact, not being able to do almost *anything* that one would like to. Most people, of course, cannot afford to do *everything* they would like to. Many cannot afford to do even *most* of what they would like to. But it is only the truly poor who cannot afford to do *nearly all* of what they would like to. For our view of poverty in Eurasia we have used just three *indicators*, as shown in Figure 10-3. The consumption (*not* production) of steel per person per year gives an idea of the quantity of manufacturing that takes place. It is manufacturing that produces goods such as

Pumping of irrigation water means that pipes have to be laid. The benefits of pipes are that the water can be taken uphill, the supply can be carefully controlled, and contamination of the water is minimized. The trouble with pipes is cost; they are dearer than open ditches.

Shell Photographic Service

A simple rice-threshing machine in Brunei. Does this solve the overpopulation problem?

Does this mean that wealth is represented by *choice*? That people are rich according to the amount of choice they have? And that whenever freedom of choice is diminished, then people are poorer?

Country	Steel consumption kg per person per year	Percentage of dwellings with indoor supply of piped water	GNP $U.S. per person per year
Afghanistan	1	5	100
Bangladesh	3	10	90
Brunei	1	88	1 500
Burma	3	10	90
China	28	10	200
Hong Kong	158	95	800
India	12	10	100
Indonesia	4	10	120
Iran	43	13	400
Iraq	48	15	350
Israel	227	90	2 000
Japan	676	90	2 000
Jordan	23	21	300
Khmer	2	10	250
Korea	80	12	300
Kuwait	196	95	5 000
Laos	3	10	100
Lebanon	99	80	600
Malaysia	36	40	350
Mongolia	1	1	100
Nepal	1	10	100
Pakistan	6	10	150
Philippines	35	30	400
Saudi Arabia	30	10	550
Sikkim	1	10	100
Singapore	150	80	1 000
Sri Lanka	8	10	180
Syria	44	20	300
Taiwan	103	25	400
Thailand	21	40	190
Turkey	30	25	450
Vietnam	12	15	250
Yemen	1	10	250
Albania	35	20	750
Austria	396	85	2 000
Belgium	477	48	2 700
Bulgaria	273	28	1 000
Cyprus	10	26	900
Czechoslovakia	611	50	1 500
Denmark	439	97	3 200
Finland	401	47	2 300
France	457	92	3 000
East Germany	520	65	1 500
West Germany	658	98	3 300
Greece	72	30	1 000
Hungary	298	37	1 200
Ireland	124	51	1 350
Italy	393	62	1 800
Luxembourg	400	95	3 000
Malta	50	72	700
Netherlands	435	90	2 500
Norway	497	93	3 100
Poland	356	47	1 200
Portugal	93	29	750
Romania	317	12	1 000
Spain	280	45	1 000
Sweden	733	95	4 100
Switzerland	474	96	3 200
U.K.	458	98	2 200
U.S.S.R.	454	50	1 500
Yugoslavia	167	30	1 000
for comparison:			
Canada	524	95	4 000
U.S.A.	620	93	5 000

Figure 10-3. Selected Standard of Living Criteria, 1971

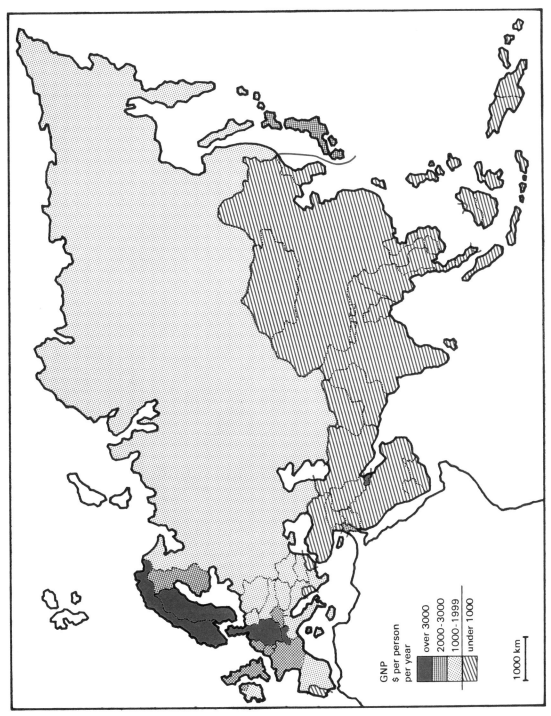

Figure 10-4. GNP in $U.S. per Person per Year

cars, washing machines, TVs, cameras, and so on. If a lot of steel is consumed per person per year, it shows that there is a plentiful production of goods. But if only a little steel is consumed, then we can safely assume that not many goods are produced either. Thus a low consumption of steel is an indicator of a degree of *material poverty.*

The second criterion listed — percentage of dwellings with an indoor supply of piped water — is another measure of poverty. Water piped indoors is something that we in Canada take for granted. Not so in much of the world. Lack of piped water leads to lack of comfort; life is harder. It may also lead to disease. Thus a low percentage of dwellings with indoor supplies of piped water is an indicator of *life-style* poverty.

The third criterion is GNP per person per year. This is a coverall criterion, since it deals with the total value of all goods and services bought and sold within a country, divided by the population. It is a rough-and-ready measure of the income that is being generated each year in each country. Like all coveralls, its fit is loose; so don't be picky over slight differences here and there. The map in Figure 10-4 illustrates the Eurasian distribution of per capita GNP. Isn't the pattern more or less what you expected? In addition to being hungry and inefficient, then, the people of south Asia are also poor, *as measured by our three criteria.*

NATURAL INCREASE IN POPULATION

Natural increase is the gap between birth rate and death rate. For example, if the birth rate in a country is 27 per 1000 (meaning that for every 1000 people in the country there are 27 babies born each year) and the death rate is 14 per 1000 (meaning that out of every 1000 people in the country 14 die each year), the gap between the two — 13 per 1000 — is the natural increase. This means that there are an *additional* 13 people each year for every 1000 who live there already. If 2000 people live in a place where the natural increase is 13/1000, then there are 26 additional people in that place at the end of the year. How many additional people are there if the N.I. is 13/1000 and the existing population is 1 000 000? That's right, 13 000. What if the existing population is 550 000 000, and the N.I. is 26/1000 — how many additional people then? That is what it is like in India: 14 300 000 *additional* people each year. More each year indeed, because the number of existing thousands already living there gets larger each year. And don't forget, that's not 14 300 000 new babies; it is 14 300 000 *additional* people, made up from the fact that the birth rate in India is 43/1000 and the death rate is 17/1000. This means that in India there are nearly 24 000 000 babies born each year, and that more than 9 000 000 people die, leaving over 14 000 000 extra people at the end of the year.

The population explosion we hear so much about is thus not caused just by the birth rate, but by the natural increase rate. Figure 10-5 shows the birth rates, death rates, and natural increase rates for the countries of Eurasia; the natural increase rates alone are plotted in Figure 10-6. The table shows that European countries generally have very low B.R. and D.R. figures; since the two figures are usually close to each other, then the N.I. gap is also, of course, very small. The populations in European countries may be quite large and the densities may be quite high, but the situation is not changing very much. In south Asia, however, things are very different. Birth rates are fairly high, and death rates are fairly low, so that the N.I. gap is quite large. Population in south Asia is therefore increasing at a fairly fast rate.

The particular danger in south Asia is that some of the countries there are already fairly crowded (e.g., Bangladesh), and the problem is just being aggravated by a large N.I. rate. The United Nations has made some estimates of how long it will take for the populations of certain countries to *double* their 1974 levels. Kuwait and Malaysia are expected to double in population by 1991; Iraq, Brunei and Jordan by 1992; Philippines, Bangladesh, Syria, and Thailand by 1993; Mongolia by 1994; Indonesia, Pakistan, Iran, and Khmer by 1995; Albania and Korea by 1996; Saudi Arabia and The Yemen by 1997; India by 1998; Turkey and Laos by 1999; and Afghanistan by 2000. By contrast, Canada's population is not expected to double from its 1974 figure of 22 500 000 until about the year 2045.

As you can see, most of the countries named by the U.N. to double their populations before the end of the century are in south Asia. If you are not sure of their locations, look at Figure 10-7. These are precisely the countries that we have already identified as *overpopulated*, according to our definition at the beginning of this chapter. They are already characterized by hunger and poverty, and their future looks bleak unless changes can be brought about.

The natural increase rate in India is adding one person to the existing population every 2 seconds. This is caused by the fact that between 45 and 50 babies are born in India every minute of the year, and that 17 to 18 people die. Thus every minute there are about 30 additional people. That's one every 2 seconds! How many is that altogether since school started this morning? How long would it take to equal the population of your community?

	Birth Rate (births/1000)	Death Rate (deaths/1000)	Natural Increase (increase/1000)
Afghanistan	50	26	24
Bangladesh	52	20	32
Brunei	39	6	33
Burma	40	17	23
China	33	15	18
Hong Kong	20	5	15
India	43	17	26
Indonesia	48	19	29
Iran	45	16	29
Iraq	49	15	34
Israel	27	7	20
Japan	19	7	12
Jordan	49	16	33
Khmer	45	16	29
Korea	39	11	28
Kuwait	43	7	36
Laos	42	17	25
Lebanon	28	5	23
Malaysia	45	10	35
Mongolia	42	11	31
Nepal	45	23	22
Pakistan	48	18	30
Philippines	45	12	32
Saudi Arabia	50	23	27
Sikkim	29	16	13
Singapore	23	5	18
Sri Lanka	30	8	22
Syria	48	15	33
Taiwan	28	5	23
Thailand	43	10	33
Turkey	40	15	25
Vietnam	37	16	21
Yemen	50	23	27
Albania	35	7	28
Austria	15	13	2
Belgium	14	12	2
Bulgaria	16	9	7
Cyprus	23	8	15
Czechoslovakia	15	11	4
Denmark	14	9	5
Finland	13	9	4
France	17	11	6
East Germany	14	14	0
West Germany	13	11	2
Greece	16	8	8
Hungary	15	12	3
Ireland	22	12	10
Italy	17	10	7
Luxembourg	13	12	1
Malta	16	9	7
Netherlands	18	8	10
Norway	17	10	7
Poland	17	8	9
Portugal	18	10	8
Romania	21	10	11
Spain	20	9	11
Sweden	14	10	4
Switzerland	16	9	7
U.K.	16	12	4
U.S.S.R.	17	8	9
Yugoslavia	18	9	9
for comparison:			
Canada	17	7	10
U.S.A.	18	9	9

Figure 10-5. Vital Statistics, 1971

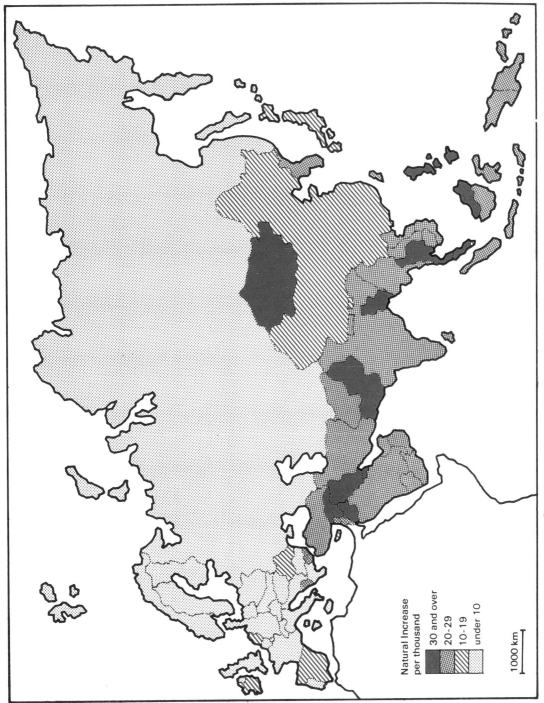

Figure 10-6.   Natural Increase Variations across Eurasia

Natural Increase
per thousand

30 and over
20-29
10-19
under 10

1000 km

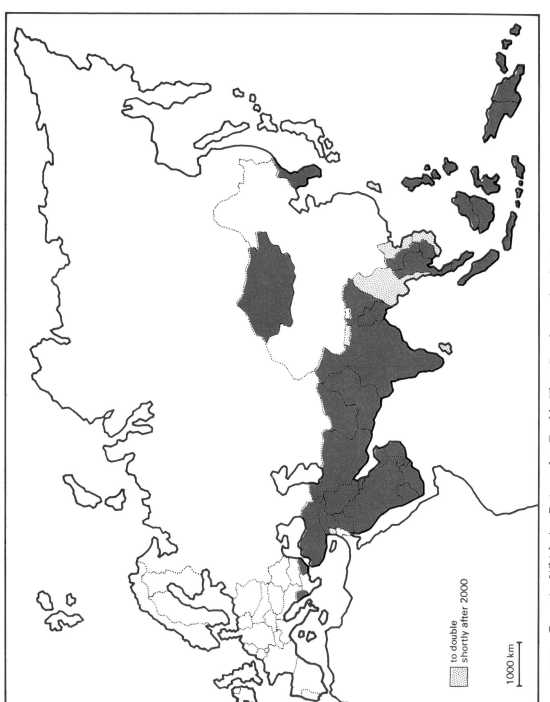

Figure 10-7. Countries Which Are Estimated to Double Their Populations by the Year 2000

to double

shortly after 2000

1000 km

All the extra people they acquire each year merely add to the burden, making it much more difficult to produce change — and yet at the same time, of course, making it much more necessary to produce change.

What can be done? The problem is so severe and so large that it must be attacked on several fronts at the same time. There is no *single* thing that can be done that will make any difference whatsoever; many things have to be done together. We cannot list them in any sort of order — there is no order. THEY MUST ALL HAPPEN TOGETHER: improved health care, better education, more factories, more housing, paved all-year roads, more fertilizer, better birth control, land reform, improved marketing arrangements, better storage facilities, increased power production, improved irrigation, more pesticides, firmer government, improved seeds, new farming methods — you name it.

You name it.

What is being done? Well, the rest of the world certainly is not letting south Asia slide to disaster without trying to help stop the slide. There are various studies of the problem, even including this one, but studies only prepare the mind; they do not actually solve the problem. They just explain the problem and get people in the frame of mind to tackle it. Then comes the tackling. There are organizations set up by the rest of the world to help south Asia, chief among which are various U.N. agencies. These agencies also help other parts of the world needing aid, but most of their help goes into south Asia. The main agencies that provide actual help in the field are the FAO (Food and Agriculture Organization), UNESCO (United Nations Educational, Scientific, and Cultural Organization), WHO (World Health Organization), World Bank, and IFC (International Finance Corporation). These various agencies do all the sorts of things that you might expect. For example, the FAO sends out farming experts to show hungry farmers how to grow new seeds, how to set up crop rotation schemes, how to deep-plough, how to irrigate, and so on. UNESCO sends out teachers and informs people of new teaching materials. WHO sends out doctors and nurses to carry out inoculations and explain about better hygiene; WHO is currently trying to eradicate smallpox. The World Bank and the IFC both help to finance major and minor projects designed to help produce more power, better transportation systems, improved irrigation, and so on. Generally the World Bank

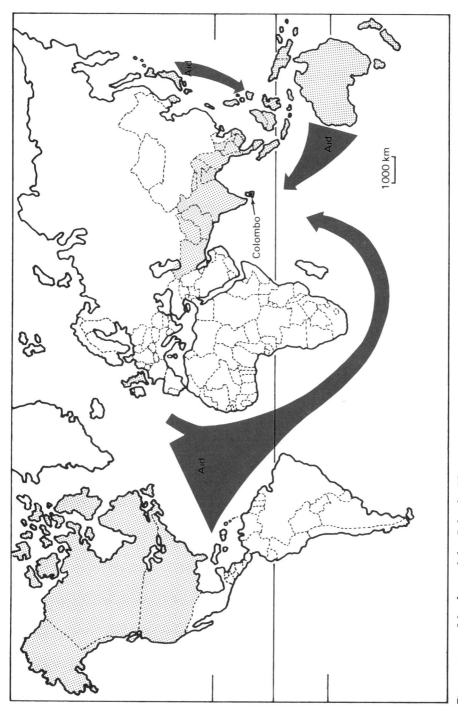

Figure 10-8.　Members of the Colombo Plan

concentrates on the larger projects and the IFC on the smaller schemes.

In addition to the various U.N. agencies there is also the Colombo Plan. The members are shown in Figure 10-8. This plan is designed to provide help to the needy countries in an overlapping sort of way, so that India, for example, receives aid in some things from the richer members, like Canada, and in turn gives aid in other things to the poorer countries, like Afghanistan. Thus India receives aid in farming technology from Canada and gives aid in educational affairs to Afghanistan.

CONCLUSION

Overpopulation is caused by the inability of a population to arrange its life-style above the level of hunger and poverty. So long as hunger and poverty are present, we can say that overpopulation also exists. The problem is severe in south Asia, and is being made worse by the high natural increase rates that characterize the region. There are possible solutions that can be applied, but great energy is needed to apply the solutions in a comprehensive manner rather than in a piecemeal fashion. Outside help is certainly available, chiefly from U.N. agencies, but also directly from richer nations.

QUESTIONS AND EXERCISES

1. What sort of work has the Colombo Plan succeeded in doing since it started in 1951?
2. What does Canada do directly to help the countries of south Asia?
3. Go to the library and research the different definitions of *overpopulation*. Have a discussion on their relative merits.
4. What are the different things that can be done in detail to help make farmers more efficient? What are the problems involved in doing these things?
5. Find out about and report on the *Green Revolution*.
6. Research into land reform in India. In particular, see what you can find out about the *Bhoodan* and *Grandan* movements.

11. ENERGY AND ECONOMIC PROGRESS

Energy may be defined as the ability to perform work. It may be produced by a great variety of things. At the very basic level, energy is produced by good nutritious food, which enables people to work well. At a slightly different level, it can be produced by animals which can be harnessed to perform work (and it should of course be noted that the animals also require nutritious food). Energy can also be produced by natural forces, in the form of water power, wind power, solar power, and tidal power. It can be produced as well from the so-called fossil fuels, such as coal, oil, and natural gas. There are sources of energy also available in atoms, chiefly those of uranium and hydrogen. And as if that were not enough, there are further supplies of energy available from the earth's interior (geothermal energy), and from the waste products of animals. An example of the latter includes cow dung in India, which accounts for about half that country's total power supply, either directly, when dried and burned, or indirectly when turned into methane gas. In the Gobi desert, on the other hand, the Mongolians have developed a portable solar oven for cooking their daily food simply by using the heat of the sun (of which there is a lot in the desert, naturally).

While there is clearly no overall shortage of energy supplies, the countries of Eurasia vary enormously in their development of these supplies. After all, even at the most basic level, good nutritious food does not just grow on trees. It has to be carefully cultivated. With different types of energy supplies, the problems of development become more complex still. Some countries can cope with these complexities of energy development on their own. Others need help. Generally, the countries that can cope on their own are also the most industrialized, the most urbanized, and the most involved in world trade. They usually have the highest standards of living, the highest GNPs, and the highest levels of technology. On the other hand, the countries that have difficulty coping with the problems of energy development also tend to be the most rural and agricultural, the most isolated in world trade. They also tend to have the lowest material standards of living and the lowest levels of technology.

Fossil fuels, as the name implies, are fuels that are really also fossils. They are the remains of things that were once alive. Trees and swamp vegetation have become coal; insects and other microscopic organisms have become oil and gas. And all these once-living things only lived, of course, by using the sunshine of those times, so we can really look at fossil fuels as being like stored sunshine.

Mick Horner

People working well in Burma.

Mick Horner

Elephants being prepared for work in the Thailand forest.

Sue Mason

A water-buffalo working in S. China.

	Population (millions)	Calories per capita per day	Electricity total mKw.h per year	Electricity mKw.h per person per year	Hydro-electricity mKw.h per year	Coal output per year 000 tonnes	Crude oil output per year 000 tonnes	GNP per capita U.S. $
China	750	2 050	n.a.	n.a.	n.a.	360 000	25 000	n.a.
India	550	1 940	60 000	109	25 000	73 000	7 000	100
U.S.S.R.	240	3 180	740 000	3 083	125 000	433 000	350 000	1 444
Indonesia	120	1 800	2 000	17	1 200	170	42 000	115
Japan	105	2 450	360 000	3 429	80 000	40 000	750	1 911
Bangladesh	75	2 000	2 000	27	—	200	—	70
W. Germany	60	2 940	240 000	4 000	18 000	112 000	7 500	3 034
Pakistan	60	2 400	5 000	83	3 000	1 000	480	175
U.K.	55	3 180	250 000	4 545	6 000	145 000	80	2 128
Italy	54	2 950	118 000	2 185	42 000	300	1 500	1 727
France	50	3 270	140 000	2 800	56 000	37 000	2 300	2 901
Korea	45	2 400	30 000	667	5 000	36 000	—	256
Vietnam	40	2 100	2 200	55	50	3 300	—	240
Philippines	38	1 990	9 000	237	2 000	50	—	377
Thailand	35	2 200	5 000	143	1 000	—	5	190
Turkey	35	2 760	8 600	245	3 000	5 000	3 500	450
Poland	33	3 140	65 000	1 970	2 000	140 000	425	600
Spain	33	2 750	56 000	1 697	28 000	11 000	150	964
Iran	29	2 030	7 000	241	1 500	300	191 000	392
Burma	28	2 010	650	23	420	12	750	90
Romania	21	3 010	35 000	1 667	3 000	6 500	13 400	800
Yugoslavia	20	3 130	26 000	1 300	15 000	650	2 900	550
E. Germany	17	3 040	68 000	4 000	1 250	1 000	—	1 500
Afghanistan	16.5	2 060	350	21	325	140	—	80
Czechoslovakia	14.5	3 030	45 000	3 103	4 000	28 000	200	1 340
Taiwan	13.8	2 620	13 500	978	3 000	4 500	90	389
Netherlands	13	3 030	41 000	3 154	—	4 500	1 900	2 353
Sri Lanka	12.5	2 210	850	68	725	—	—	180
Nepal	11	2 030	110	10	40	—	—	70
Malaysia	10.5	2 200	3 500	333	1 200	—	850	350
for comparison:								
Canada	21	3 200	210 000	10 000	160 000	12 000	60 000	3 676

Figure 11-1.   Energy Data for the 30 Most Populous Countries in Eurasia, 1970

Figure 11-1 contains some data on energy supplies available on a per person basis for the 30 most populous countries in Eurasia. You can see that the countries that rank highly in one characteristic also tend to rank highly in other characteristics. Figures 11-2 and 11-3 show another way of looking at some of these data. Figure 11-2 is a scattergraph of the relationship between per capita annual electricity supplies and per capita daily food intake measured in Calories. You can see quite clearly that there is a positive correlation; as Calorie intake rises so does electricity output. Needless to say this does not necessarily mean that people produce more electricity *because* they eat more food, or that they eat more food *because* they produce more electricity. However, it does indicate that *something* is at work causing people to have more food as well as more electricity in some countries, and less in others.

Hydro dam at Laggan, Scotland.
British Information Services

Figure 11-3 shows the correlation between electricity supplies and per capita GNP, and again you notice a strong positive correlation. Could it be, do you think, that if you eat nutritious food you will be able to generate a

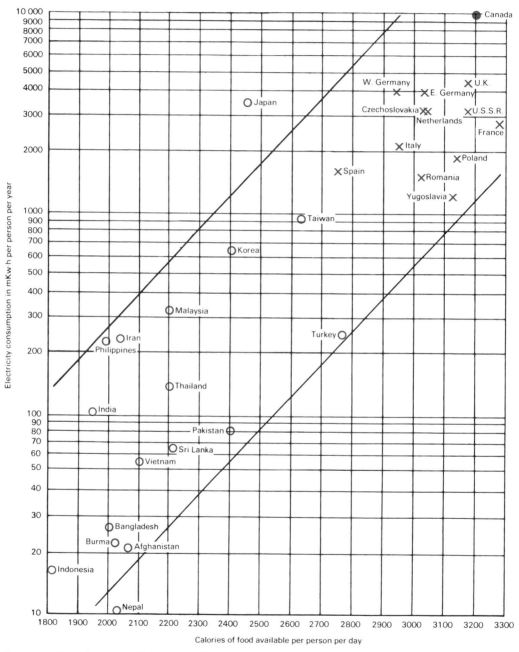

Figure 11-2.   Electricity/Calories Scattergraph

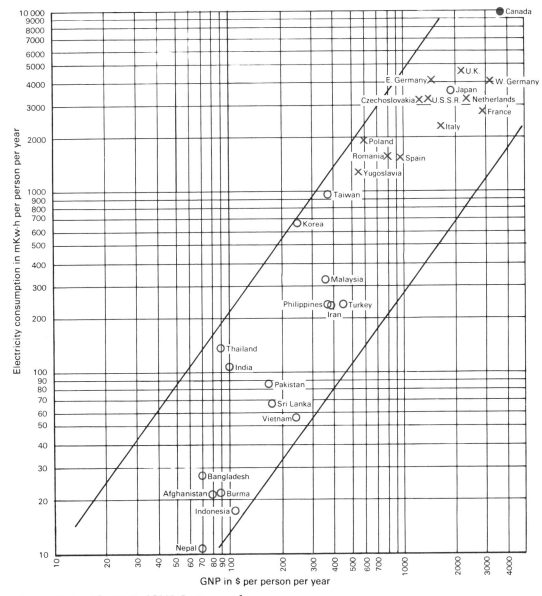

Figure 11-3. Electricity/GNP Scattergraph

lot of electricity and thus become rich, or is it more likely that if you generate a lot of electricity you will grow rich and be able to eat more? Or is it perhaps that if you are rich to start with, then you can afford to generate a lot of electricity and eat more as well? What do you think? Perhaps something else entirely enables all these things to happen together (education maybe?).

We should now be readily prepared to accept that energy supplies are very important to a country's development. So let's have a look at some of the major supplies that already exist in Eurasia.

Coal, oil, and hydro-electricity are undoubtedly the big 3 in the energy field, and most of the major countries in Eurasia have access to at least one or other of them. Some countries, notably the U.S.S.R. and China, are fortunate enough to have supplies of all 3, as well as supplies of nuclear power. Other countries, notably Japan and West Germany, cannot produce anywhere near enough power for all their needs, and so they are extremely reliant on energy imported from elsewhere. As a result of these shortages in certain countries and surpluses in others, there is a considerable trade in energy products (chiefly oil).

The map in Figure 11-4 shows the national variations in annual coal production throughout Eurasia. Three giant concentrations of output are apparent:

1 Europe, chiefly the U.K., Poland, and West Germany;
2 The U.S.S.R., chiefly the Donbas, Kuzbas, and Kara-ganda regions;
3 China, especially the provinces of Szechwan, Shensi, Shansi, and Manchuria.

In Europe, the U.K. produces coal from a number of individual scattered coalfields rather than from just one or two large fields. The coal is of good quality, and is widely used for a great variety of purposes — both indirectly, as in steel making, and directly, as in thermal electricity generation. The chief problem facing many of the smaller coalfields in Britain is that the individual coal seams are often very thin (1 m or less) and usually quite twisted. Figure 11-5 gives a simplified geological section of a British coalfield. You can see that the exposed part (which was the easiest to mine, and usually the first part to be used) is now suffering some decline because all the best coal has already been taken out. The industry in the concealed parts, however, is thriving, and is likely to continue to do so just so long as coal is needed. There are

Oil and gas are the most-traded of all fuels partly because they are so easily and cheaply transportable compared to the other fuels and partly because the largest areas of production have only a small industrial demand, thereby leaving huge potential surpluses for export (provided the governments are willing to let the oil flow freely).

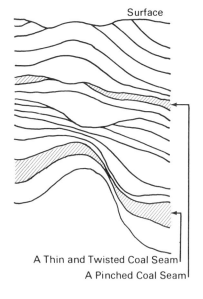

Surface

A Thin and Twisted Coal Seam
A Pinched Coal Seam

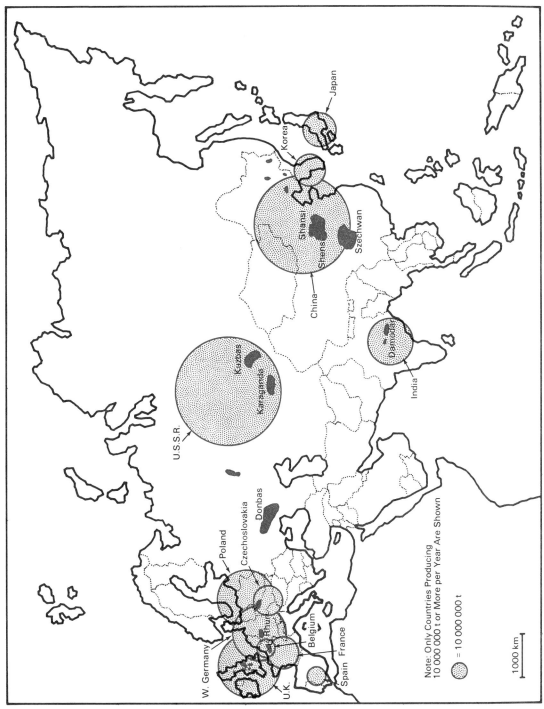

Note: Only Countries Producing
10 000 000 t or More per Year Are Shown

⬤ = 10 000 000 t

1000 km

Figure 11-4. Major Coalfields and Relative National Outputs

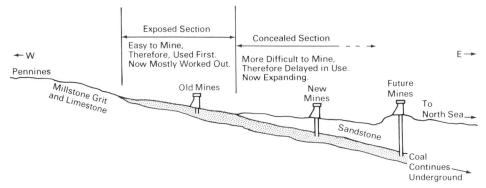

Figure 11-5. Simplified Section through the East Pennine Coalfield

official reserves in the concealed parts that will last at least another 800 years at current rates of use.

Poland (chiefly the district of Silesia) produces almost as much coal as the U.K., and it has the added advantage that the seams are very much thicker (usually more than 1 m, and in places as thick as 10 m). Again the known reserves are huge, and will last hundreds of years. In West Germany, the Ruhr is one of the world's outstanding coal and industrial areas. The coal is produced under conditions very similar to those in Britain (i.e., the exposed parts of the field are experiencing declining output, while more modern production is becoming concentrated in the concealed parts). The quality is also exceptionally high, and the official reserves are sufficient to last perhaps another 1500 years at current rates of use.

If you refer back to Figure 11-4 you can see that coal production in the U.S.S.R. is also scattered throughout a number of separate fields. By far the largest, however, are the DONetz BASin (DONBAS) field just north of the Black Sea, the KUZnetsk BASin (KUZBAS) field in central Siberia, and the Karaganda field, also in central Siberia. The remoteness of the Siberian fields from the main population centres in western Russia poses a problem, but the Moscow government is trying to cope with it by developing heavy industry in Siberia near the coalfields. If the coal itself had to be moved into western Russia, then it has been estimated that as much as half the coal would be used up in supplying enough energy to transport the other half.

Coal is also scattered in China, which tends to suit the current Chinese philosophy of self-sufficiency in the communes. The main deposits are shown in Figure 11-4, where

Britain's coal output reached its peak in 1913, when 217 000 000 t were mined. Since then output has declined. Latest figures are:

1966	186 000 000 t
1968	174 000 000 t
1970	163 000 000 t
1972	154 000 000 t
1974	145 000 000 t

Why do you think this decline has occurred?

it is clear that the Shensi-Shansi field and the Szechwan field are the most important. The largest individual coal workings occur, however, in the Manchurian fields (there are many scattered deposits), where, at Fushun, the thickest known bed of coal in the world is worked. The seam is about 125 m thick, and is worked in a giant open-pit mine. You will not find Fushun marked on Figure 11-4; the map is too small. Look it up in your atlas.

The only significant deposits of coal in Eurasia outside the 3 concentrations already mentioned are in India, where 98% of the useful coal lies just west of Calcutta along the Damodar valley. Mining techniques are not so mechanized as those in Europe or the U.S.S.R., and the production of coal per worker is therefore very much less. Nevertheless, the quality of the coal is very good, and it supports important steel and engineering industries in the Damodar valley region.

Compared with the distribution of coal workings, oil production is much more concentrated into only a few areas, as you can see in Figure 11-6. One of the reasons for this is that it is very much more difficult to find oil deposits, and if it hasn't been found yet, then it cannot be produced yet, either. The Russians, for example, are right now carrying on an intensive oil-search in eastern Siberia; the Chinese are investigating their continental shelf in the South China Sea; and the Europeans are just beginning to bring North Sea oil ashore after having discovered its existence only during the early 1970s. Another reason for the concentration of oil production into only a few areas is that oil is by far the cheapest form of energy to transport over long distances. This means that it can be produced in one area and used in another, which may be thousands of kilometres away, without there being a significant price difference. It's often therefore cheaper to bring it in from a known oil area than to search for it in your own backyard. A third reason for the concentration of production, especially in the Middle East, is that production costs in some areas are so low that the major oil companies (the so-called "Seven Sisters"— can you name them?) and oil countries find it cheaper to produce there and transport the product around the world than to search for it in areas where production is likely to be more expensive. For example, the production costs for a barrel of oil are about a dime in the Middle East and about a dollar in Alberta. The difference between the production cost

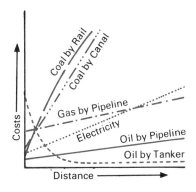

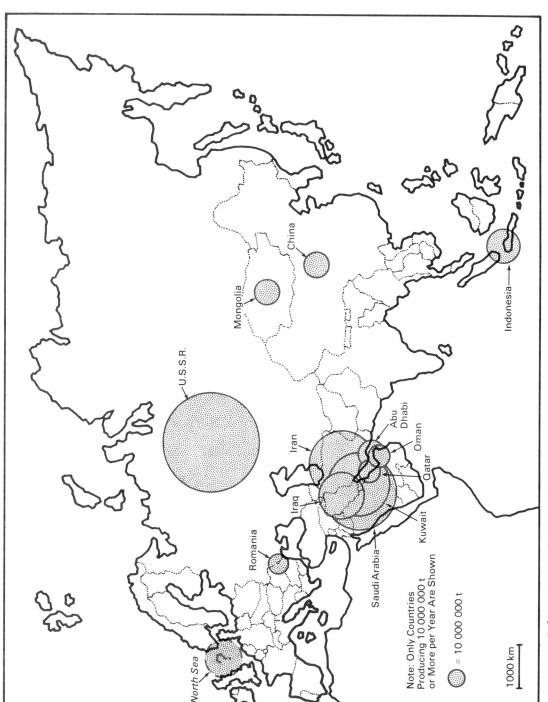

Note: Only Countries
Producing 10 000 000 t
or More per Year Are Shown

= 10 000 000 t

1000 km

U.S.S.R.

Mongolia

China

Indonesia

Iran

Abu
Dhabi

Oman

Qatar

Iraq

Kuwait

Saudi Arabia

Romania

North Sea

Figure 11-6.   Relative National Oil Production

and the selling price is made up of taxes, royalties, and profits, mostly taxes and royalties.

In 1973-74, however, this pattern of concentrated production began to change very quickly. The Middle East producers suddenly (and sharply) raised their royalty demands, making oil much more expensive for other countries to buy. Many other countries had already started to develop their oil industries, of course, because they did not want to rely solely on foreign supplies anyway, but the raised royalties speeded up the process of development. Oil production began to be much more scattered, and generally also much more expensive, because oil-searching had to be done in less favourable areas — the most favourable already having been developed — and production had to take place from the more difficult fields.

The Middle East is likely to remain the major producing area, however, no matter what happens in other parts of the world. Production costs are so low in relation to those of other areas, and the oil is so easy to obtain, that the Middle East countries will always be able to undercut the prices charged at other fields if they should want to. In any event, the known reserves of oil in the Middle East are so great that production is guaranteed for many decades. It is a measure of how expensive oil production is outside the Middle East (and also of how few competitors there are) that the Middle East countries were able so successfully to raise their royalty rates suddenly in 1973-74.

The U.S.S.R. has long had interests in the oil business, starting with the development of Baku in the 1870s. The Baku field still produces oil, but costs are rising rapidly as the field becomes exhausted. During the 1930s replace-

An oilfield grows old. When the field is young, the oil gushes under its own pressure, and it has to be controlled by caps and valves. As the field grows older, the oil ceases to flow under its own pressure, and it has to be pumped out, thus raising the costs of operation. As the field nears its end, the oil can no longer even be pumped out, because it is so spread out among the pores in the rock. It can then only be obtained by having water pumped down from the surface; since oil floats on water, the oil is driven towards a pumpable concentration near the bottom of the well; the costs naturally rise still more. The field eventually becomes exhausted.

ment oil was discovered in the Volga-Urals field, whence oil is now pumped 1500 km into the industrial regions around Moscow. Additional supplies have recently been found in the Tyumen district, and Russian oilmen are now confident that Siberia has vast treasures still to be discovered. The combination of supply sources is sufficient to rank the U.S.S.R. after the U.S.A. in individual national oil production.

Europe's oil production was for a long time concentrated in the relatively tiny Ploesti fields in Romania, but World War II forced changes to occur. Intensive searches for oil during the war revealed natural gas supplies only. The search was continued after the war, revealing more gas fields (in the northern Netherlands). For a while it seemed that gas was going to be the only product from all the searching. But as oil technology improved and it became possible to search far out under the waters of the North Sea, the first liquid oil supplies began to be found. In the late 1960s and early 1970s oil was discovered in large quantities at a number of separate fields under the North Sea. The Europeans, especially the British, hope to have supplies coming ashore in large and reliable quantities by the 1980s.

Coal and oil are examples of what are called *primary fuels*. Most electricity is a *secondary* power source. Electricity may be made in a variety of ways, such as by using the force of falling water to drive turbines (called hydro-electricity), or by burning coal or oil to boil water to produce steam to drive turbines (called thermal electricity), or by harnessing and controlling the power of atoms (called nuclear electricity), or by such other methods as focusing the sun's energy or using the power of the tides. The total amount of electricity produced by some or all of these methods together is shown in Figure 11-1; so is the amount of hydro-electricity. Hydro-electricity is different from thermal electricity only in its mode of production, but it is an important difference because it represents a completely new addition to the supply of power in a country. After all, thermal electricity is only processed coal and oil, and we shouldn't really count them twice, should we? Nuclear and tidal electricity also represent new additions to the power supply, in the manner of hydro-electricity, but at the moment they add only a very small proportion to the supply, so they are not shown separately in Figure 11-1.

The production of hydro-electricity throughout Eurasia

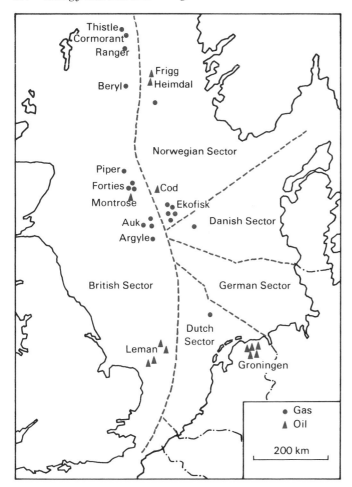

Figure 11-7.   North Sea Oil and Gas Discoveries

varies according to the extent of availability of a number
of contributory factors:

1   *Regular flow of water*, which in turn depends upon
    such factors as seasonal rainfall distribution, winter
    freeze-up, summer evaporation loss, permeability of
    rock in the drainage basin, the existence of lakes along
    the river, and land use in the drainage basin (lots of
    buildings help speed up the water run-off after rain,
    while extensive forests hold back the water after rain,
    thus helping to even the flow).

2   *Large volume of water*, which depends directly upon
    the amount of precipitation during the year. Lakes and
    reservoirs do nothing at all to increase the volume of
    water, because you can only take out as much water
    as the precipitation puts back in. The purpose of lakes

Permeability is a measure of
how well rock lets liquids pass
through it, either by way of
cracks (fissures) or by way of
holes (pores). If liquids pass
through easily, the rock is said
to be *permeable*, but if liquids
do not pass through easily (perhaps even not at all) then the
rock is said to be *impermeable*.

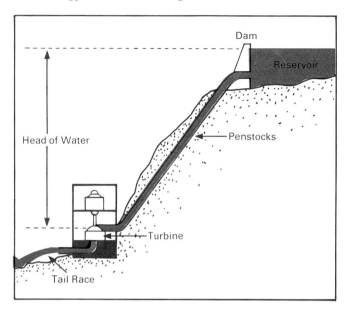

and reservoirs is to regulate the flow, so that run-off
is not wasted during times of heavy rainfall but is
reserved instead for supply during periods of low
precipitation.

3 *Large head of water*, which varies according to the
topography. The head of water is the difference in
height between the top of the supplying water surface
(usually a reservoir behind a dam) and the bottom of
the penstocks feeding the water into the turbines. The
turbines are not necessarily located at the base of the
dam. They can be located anywhere, even on the other
side of a mountain range, just so long as the maximum
head of water is obtained by the arrangement.

4 *Proximity — being close — to a market*, which may
be a single industry or a large population cluster.
Generally electricity is the most expensive form of
power to transport over long distances, even with the
new ultra-high-voltage lines. Thus the market should
be within approximately 500 to 1000 km of the supply
centre, preferably even closer than that. The exact
distance over which it is *economic* to transport electricity depends, of course, on the availability and cost
of possible competing fuels. However, there must be
a market for hydro-electricity somewhere close to the
supply, otherwise the physical factors, although ideal,
will not be developed. Indeed, if the market is a long

way away, it might pay the market to import coal or oil and generate thermal electricity instead.

The map in Figure 11-8 shows the variations in hydro-electricity production across Eurasia. Four concentrations stand out: Scandinavia, Western Europe, the U.S.S.R., and Japan. All these areas, of course, provide markets for power in one way or another. In Western Europe, the U.S.S.R., and Japan the markets are a direct result of the industrial demands of large domestic populations, but in Scandinavia the markets are slightly different. Domestic populations are quite small (none of the Scandinavian countries are in the top 30 shown in Figure 11-1), but the other conditions are so good and the area in general so accessible by ships from other parts of the world that hydro-electricity has reached a high stage of development. The market for much of the electricity produced in Scandinavia is, in fact, the processing industries set up to supply the industrial needs of other countries. For example, raw ores are shipped in, processed, and shipped out again in some fabricated form. The electro-metallurgical and electro-chemical industries are accordingly quite important in Scandinavia. An additional reason for the high degree of hydro development in Scandinavia is the lack of any key competition. There are no major coal supplies, no oil or gas (except for some that Norway has recently acquired in the North Sea bonanza), and no nearby surplus suppliers of any of these primary fuels.

It is the absence of major competition that has also led to the large-scale development of hydro power in France. France, with paltry supplies of coal, was left behind in the early years of coal-based industrialization. It accordingly determined that it would develop all available supplies of alternative fuels as soon as the technology was invented. As a result, France has avidly sought oil and gas in its own territory (successfully in the south-west). It has hungrily pioneered the development of solar power (in the Pyrenees) and the manufacture of tidal power (at the mouth of the Rance River in Brittany). And it has gone to great trouble to develop its full hydro-electric potential. The chief developments are in the Alps, the Central Plateau, the Pyrenees, and along the Rhone River.

The U.S.S.R. has some of the very largest hydro schemes in the world (rivalling Canada's Churchill Falls and James Bay projects). Part of the reason for this is the reliably huge volumes of water in the rivers coming down from the backs of the great mountain system in central

The main research centre into solar power is at Mont Louis in the eastern Pyrenees. At Mont Louis there is a powerful solar oven, where the heat from the sun's rays is focused by a set of large reflectors (a little like the heat which can be focused by a magnifying glass). The solar oven can generate temperatures high enough to melt steel. Unfortunately the cost is extremely high, and the oven only works when the sun is shining. The fact that the sun never shines at night means that storage of electric power becomes another problem. Solar power works, but it's not without its problems.

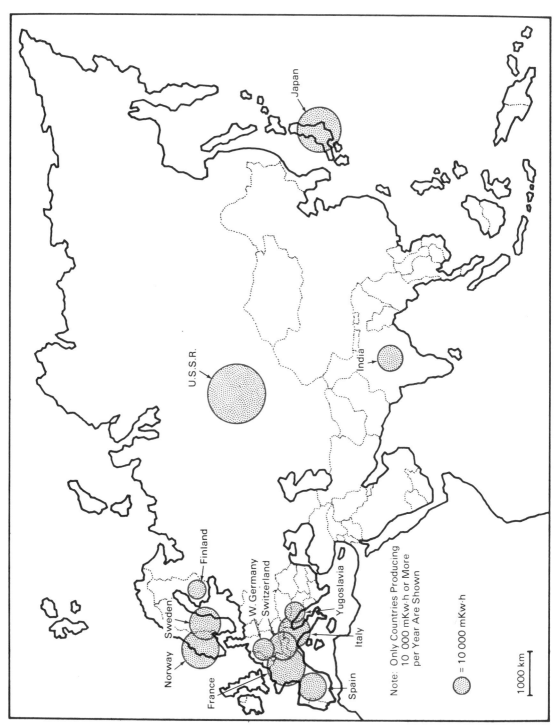

Figure 11-8. Relative National Hydro-electric Production

Japan

U.S.S.R.

India

Finland

Norway

Sweden

W. Germany

Switzerland

France

Yugoslavia

Italy

Spain

Note: Only Countries Producing
10 000 mKw·h or More
per Year Are Shown

⬤ = 10 000 mKw·h

1000 km

Asia. The developments on the Angara-Yenisei River system deserve special mention, especially as the outputs from each of its major power sites — Sayanskaya, Krasnoyarsk, Yeniseyskaya, Sukhovskaya, Bratsk, and Boguchany — are all individually among the largest in the world.

Japan has no large rivers, but it has a lot of hydro-power sites. Lack of competing fuels, plenty of rain (from the monsoons), many mountain lakes (in extinct volcanic craters), and a large industrial and domestic demand all combine to favour the development of hydro-electricity in Japan. So great is the demand for power that Japan has built over 1000 hydro plants, although most of them are quite small; the largest ones are in central Honshu, near Tokyo, where sites have been developed on the Shinano, Tone, Agano, Kiso, and Tenryu Rivers.

During the early 1970s many people throughout the world began to complain of what they called an energy crisis. They forecast that the world was soon going to run out of energy. As it stands, of course, that is pure nonsense, because there is no way — even at worst — that the world will run out of water for hydro-electricity. What *is* going to happen is that the world will eventually run out of fossil fuels; coal, oil, and gas will eventually all be gone. However, "eventually" is a long time off in the future, perhaps hundreds of years. Even when conventional oil and gas reserves are all used up, it will still be possible to produce oil and gas from coal, from oil sands, or from oil shales. Worrying about the disappearance of fossil fuels now is rather like people in the middle ages in Europe worrying about the disappearance of bows and arrows and suits of armour. In any event fossil fuels are likely to be increasingly replaced as sources of energy by renewable or breeder sources in the near future. Nuclear power is already being developed on a production (rather than experimental) basis; the share of Eurasian countries in production is shown in Figure 11-9. The dominance of European countries is quite obvious, although they are currently running into various development problems (note the 1974 decision by Britain, the largest nuclear producer, to buy nuclear expertise from Canada and the U.S.A.). Some of the biggest problems are concerned with the quality of the operating technology and the slow development of the breeder reactor (which will breed its own fuel!). Canada has a top reputation in its development of operating technology, and

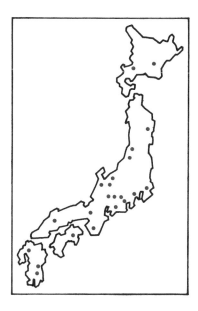

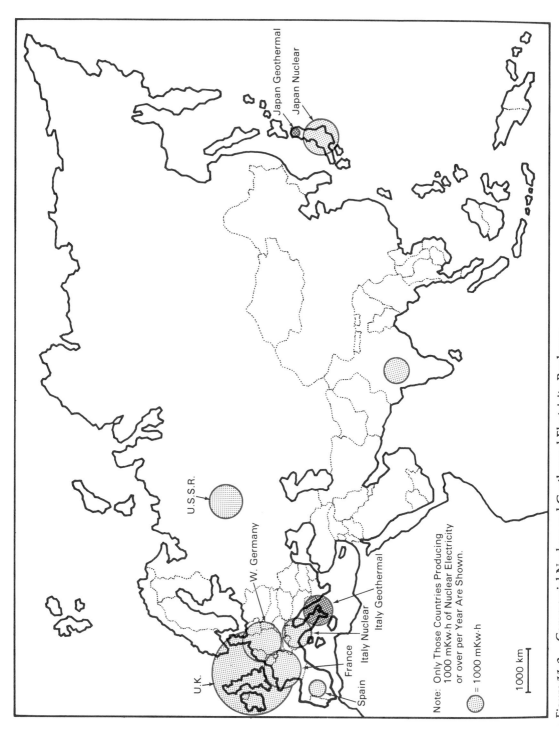

Figure 11-9. Commercial Nuclear and Geothermal Electricity Producers

various Eurasian nations (chiefly India, Korea, and Romania) have bought the Canadian CANDU system. The prospects for the continued growth of energy supplies and the development of higher levels of industry are clearly favourable, which is a satisfying thing for all the countries in the world that still do not have even electricity supplies for all their people. The only possible problem that could arise would be if lots of countries were to start using their nuclear electricity-generating capacity to make atomic bombs instead. It was a sad time for many in the world in May 1974 when India used its Canadian-built reactors to produce the fuel to set off its first atomic bomb, despite agreements with Canada that the nuclear capacity would be used for peaceful purposes only.

Netherlands National Tourist Office
An old windmill in Holland.

Krasnoyarsk hydro station on the Yenisei river.

Consulate General of Federal Republic of Germany

QUESTIONS AND EXERCISES

1. Construct a scattergraph of GNP against Calories for the countries listed in Figure 11-1.
2. What are the different *uses* of energy? Write an essay.
3. For European countries *as a whole*, including the U.S.S.R., and for Asian countries *as a whole*, calculate:
   - (i) average per capita GNP;
   - (ii) average per capita electricity production in mKw.h;
   - (iii) average per capita food available in Calories.

   What conclusions do you draw?

   Have you any explanations?
4. Some of the Middle East oil producers have reserves to last hundreds of years; others have reserves to last only a few decades. What do you think their policies will be regarding sales and prices? The same? Different? If different, then how? Do you think OPEC has a future?
5. What advantages do you think electricity possesses over other types of fuel? Do the other types of fuel possess any advantages over electricity?

12. THE ARAB WORLD AND PETROLEUM

CONSTANT CHANGE

The Arab world is a large one, occupying a great part of the area known as the Middle East, and extending along the north shore of Africa as far as the Atlantic Ocean. In this study we shall only be concerned with those Arab countries that are in Eurasia — and of those, especially the ones that contain reserves of petroleum. "Petroleum" means literally "rock oil", coming as it does from the Greek word *petra*, a rock, and the Latin word *oleum*, oil. The term "petroleum" covers a wide range of naturally occurring minerals, from asphalt, bitumen, pitch, and tar to crude oil and even natural gas, although gas is usually referred to as distinct from petroleum.

The Arab world itself has a common religion, centring around the worship of Allah and reverence to his prophet Mohammed, who lived about 1000 years ago. Arabs are a group of people of distinctive ethnic characteristics, descended from a common stock. According to the *Concise Oxford Dictionary*, this is sufficient to describe the Arabs as a *race*.

The area shown on the map in Figure 12-1 has been in a state of flux for centuries. Even before the Arabs arrived, migrating peoples had wandered over the vast spaces of this area, pasturing their flocks and herds, fighting now and then to preserve their rights of access to water and pasture, or succumbing to the greater might of established empires like that of the Babylonians. It was in this part of the world that people first learned how to support life in cities by domesticating animals and plants, farming them in fixed locations that were taxed by the great ruling cities. The history of the city of Jericho, for example, goes back fully 9000 years. In those days, however, the region was much wetter and cooler than it is today. Grassland and forest were widespread, and the land supported a large number of nomadic herders and their families, as well as a great many small farming communities. As the centuries wore on, the climate became drier. The rainfall decreased until much of the area had less than 250 mm of precipitation annually. The area became a desert.

The onset of desert conditions made life harder for the

No-one knows exactly how oil was formed, but a lot of people believe that it is the remains of tiny sea creatures that lived millions of years ago. They died and their remains decayed under special conditions on the deep-sea floor to form oil. Over the millions of years, the oil gradually collected in vast pools. The ones that have been discovered and exploited are our oilfields of today.

In parts of the Arabian desert, it has not rained in living memory. Without water, a man would last less than 2 days, for as his body perspired in an effort to keep cool, it would lose water. As a result, the blood would thicken so much that it could no longer be pumped around the body, resulting, obviously, in death.

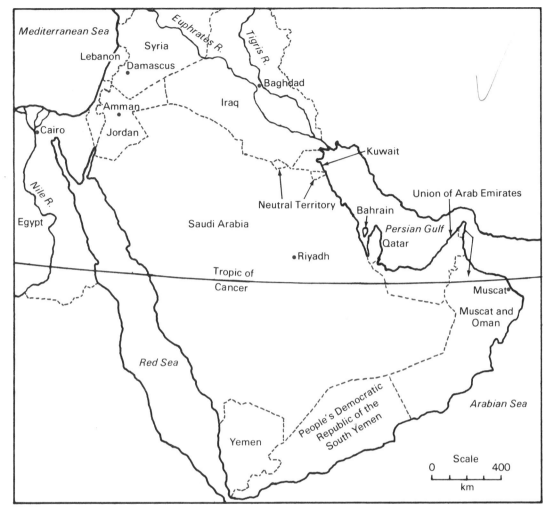

Figure 12-1. The Arab World of the Middle East

inhabitants, and many of them migrated clear away to more congenial regions. The ones who stayed behind in fixed settlements relied almost entirely upon irrigation for their crops. They tapped the water in the ground beneath their land or channeled it from rivers such as the Tigris and Euphrates, or the Nile. When their water supply failed, they had to move or die.

Over-irrigation of the land for many years made it saline or alkaline, because irrigation water often contains dissolved minerals that are deposited when the water evaporates under the fierce heat of the desert sun. Large parts of the fertile crescent, which had extended from Mesopotamia (the land between the rivers Tigris and Euphrates) into the Nile Valley, became desert. Civiliz-

Shell Photographic Service

The desert interior of Muscat and Oman.

ation dwindled as the sands of the desert crept across the land. Only the hardy Bedouin or oasis people could survive such harsh conditions. Thus the Middle East became a backwater in a world that was rapidly becoming a foaming torrent of technological change and discovery, exploration and exploitation. No-one seemed interested in the region, for it seemed to have nothing to offer. Over 400 years ago it became part of the Turkish Ottoman Empire, but still no-one was interested in it, until about the turn of the 19th century.

	J	F	M	A	M	J	J	A	S	O	N	D	
Baghdad	9	12	16	22	27	32	35	35	31	27	17	2	average temp. (°C)
	30	33	33	23	5	—	—	—	—	3	20	30	precipitation (mm)
Aden	24	24	26	28	31	32	31	31	31	29	27	25	average temp. (°C)
	8	5	13	5	3	3	—	3	3	3	3	3	precipitation (mm)

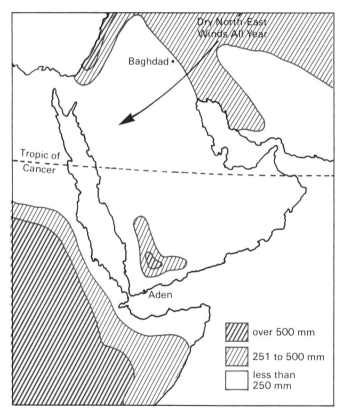

Figure 12-2.   The Annual Precipitation of the Arab World

Climate figures such as the monthly averages given above tell their own story. The region could not support very many people by farming, nor was it ever likely to.

It hardly seemed worthwhile for anyone to make large investments in transportation routes such as railways when there was likely to be little economic profit or none at all. The few railways that were built were in Iraq and Jordan and Saudi Arabia. They were the product of German foreign aid in the years before the First World War when Turkey, at the head of the Ottoman Empire, was allied with Germany. The railways were therefore built for political reasons and were used by the Turks as a means of keeping a close control on the Arabian portion of their empire.

When Gottlieb Daimler of Germany perfected the internal combustion engine in 1885, few people associated it with the desert of the Middle East. But the Middle East was rich in the fuel needed to power this engine. Let us examine the countries of the Middle East and try to assess the impact that oil discovery and exploitation has had upon them. Bear in mind that oil was first discovered in the region in 1901 in Iran, which is the only non-Arab country in the region, even though the national religion is Mohammedanism.

Before the wide-spread use of the internal-combustion engine, oil in the Middle East was simply refined for fuel oil for ships, as well as for grease and a few other things. The gasoline that was produced was mostly allowed to evaporate as a waste product!

THE UNION OF ARAB EMIRATES

Originally these were known as the Trucial States. If your atlas is an old one, then it will certainly be labelled by the latter name. Historically, these 7 small Arab states, each ruled by independent shiekhs, were given to perpetual raiding, warfare, and piracy. Why they were forced to sign a truce limiting their aggression only to one another and by land attacks only, is rooted in history. The Arabs of the region were given to preying on British shipping. The Persian Gulf is on the main sea route to India, and the Arabs could not resist the tempting booty that a ship of the East India Company could provide. The truce they signed in 1853 was maintained by force by the British administration in India. The countries involved in the truce are shown in Figure 12-3, and their names are underlined. Recently, Qatar and Bahrain have become associate members of the U.A.E.

Before the discovery of oil in the Trucial States, the inhabitants had been dependent on a little farming, using scarce water for irrigation; trade, including a small slave trade; pearling and fishing. Pearling, the only source of cash revenue, was dealt a mortal blow by the Japanese in the 1930s, when they invented the process of culturing

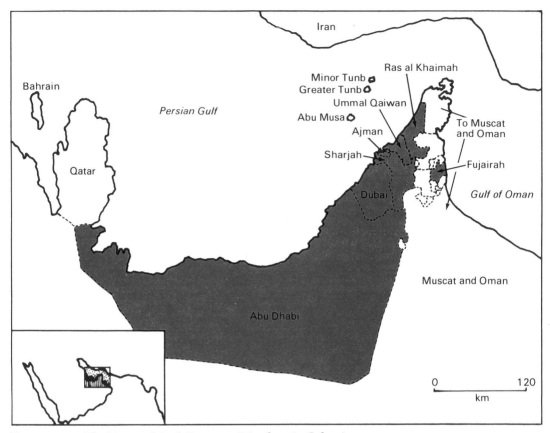

Figure 12-3.   The Union of Arab Emirates (Members in Colour)

People in the country now known as Muscat and Oman, as well as many Gulf States, were major participants in the slave trade. *Dhows* would sail down to Zanzibar, where they would pick up Africans who had been taken by force and marched under appalling conditions to the coast from as far inland as Lake Victoria. It is estimated that as many as 12 000 000 slaves were taken by Arab traders in the eastern part of Africa. Packed into the dhows in scores and hundreds, the unfortunate survivors of the coastal trek died in droves. Mortality rates of over 90% were not uncommon on the 12-day voyage north to the slave markets. The British put a stop to most of this activity between 1822 and 1874, but it has never been completely stamped out.

pearls. Disaster seemed imminent for these states and would have been were it not for the discovery of oil. It was discovered first in Bahrain and, encouraged by success there, the Petroleum Development Trucial States

Ltd. began buying the rights to explore for and develop oil. *Abu Dhabi* sold its rights in 1953, oil was discovered in 1958, and the first oil was exported in 1962. *Dubai* sold its rights in 1952, oil was discovered in 1966, and first exported in 1969. *Sharjah* first discovered oil in 1972 and exported its first crude in 1974. So far, none of the other Emirates, as they are now called, has produced oil.

The impact of oil on the U.A.E. is just starting. Each of the Emirates that has oil revenues is pledged to allocate 10% of it to finance development in the other states, who are, meanwhile, looking into the possibility of discovering oil themselves. So far the most likely candidate is *Ras al Khaimah*.

Oil has greatly changed the traditional pattern of society in the U.A.E. Allegiance used to be from an individual to his tribe. Nowadays, allegiance is paid more to a state, and national boundaries have been fixed on the map. However, because the tribes of each state were scattered in 1952 when the first political map of the area was drawn, several states are not *contiguous entities*. This means that they are in separate pieces, much as Pakistan used to be. Today, only Abu Dhabi, Dubai, and Umm al Qaiwan are contiguous territories in the U.A.E.

After the British Foreign Service had fixed the boundaries to the satisfaction of most people, it only remained for the first census to be conducted in the U.A.E. The results of this 1968 census are in Figure 12-4. It is not easy

	Population		Land area (km²)
	1968	1970 (est.)	
Abu Dhabi	46 500	60 000	66 560
Ajman	4 200	5 500	256
Dubai	59 000	75 000	3 840
Fujairah	9 700	10 000	1 152
Ras al Khaimah	24 500	30 000	1 664
Sharjah	31 500	40 000	2 560
Umm al Qaiwan	3 700	4 500	768
†Bahrain	182 000	205 000	665
†Qatar	*	80 000	10 240

*Qatar did not partake in the 1968 census.

†In 1968, Britain announced its intention to leave the region in 1971. Accordingly, the Emirs of the Trucial States and the Emirs of Bahrain and Qatar met in Dubai and announced the formation of the Union of Arab Emirates, totalling 9 states, effective December 2, 1971.

Figure 12-4.   The Populations and Land Areas of the States of the U.A.E.

for the U.A.E. to conduct censuses, because part of the population is nomadic and constantly crossing state boundaries. Thus it is difficult to assign these people to one state or another. Furthermore, and most significantly, there is a high rate of illegal immigration from other Arab countries. The illegal immigrants have come for obvious reasons: to try for a share in the new prosperity of the region.

In the early 1950s, all present 9 members of the U.A.E. had about the same level of development, or rather, undevelopment. Their towns lacked facilities such as water, sewage, electricity, paved roads, airports, hospitals, clinics, and schools. Today Bahrain has the most advanced system of communications in the Middle East. Although it sold its oil concessions in 1928 and discovered oil in 1932, with the first shipment in 1934, it had long been a trading centre and as early as 1920 had the funds to start an impressive program in higher education. Today, oil revenues have boosted this start dramatically. There are numerous telephones, a piped water supply and a modern harbour. There is also a large aluminum smelter, using local natural gas, and a flour mill.

Qatar first sold oil concessions in 1935, discovered oil in 1939, and exported its first crude oil in 1949. Unlike Bahrain, which had profited from pearling and trade before its oil boom, Qatar was a sleepy little fishing region. In 20 years Doha, the capital, has been transformed. It is a thorough-going modern city with a phenomenal expansion in commercial activities. These include a modern harbour at Doha, two water-desalting plants, modern fish canneries, and a factory producing cement. Modern architecture abounds.

Dubai has the only deepwater port along that part of the coast. Most of the oil revenue that has poured into this little country in the last few years has been used to further improve port facilities. Air-conditioned highrises have sprung up along the waterfront, and there is a fine new airport. The old markets are still there, but nowadays are most likely to be selling Swiss watches and Japanese cameras. A new tunnel for cars is planned to replace the old wooden ferry that plies across the tidal river that splits Dubai in two. Many merchants are now millionaires, for trade has grown 700% since the oil boom began in 1966. Most credit for the prosperity of Dubai must surely go to Sheikh Rashid ibn Said al-Makhtoum, the ruler of the country. He had begun to modernize the country in the 1950s, building a road network, reclaiming land for building, and he had even set up an electricity company by 1961. His oil revenues are planned to provide the largest tanker dry-dock in the region between Europe and the Pacific. Additionally, he has recruited many competent administrators from many countries and walks of life, including Jack Briggs from England to head Dubai's police force. At least 13 foreign banks have opened branches in Dubai, including the First National City Bank of New York.

Abu Dhabi has its capital on a small island offshore. Its taxi drivers never seem to know their way around it, because new buildings are completed every month! Only a few years ago, the camel-herding nomads or pearl fishers were poor. Today, they have free houses, free medical care, and free education through to university. Although all consumer and capital goods must be imported, there is plenty of money to pay for them. In 1969-1970, the demand for electricity rose by more than 10 times, the biggest jump anywhere at any time in the world. Plans are being implemented for an oil refinery, a petro-chemical plant, and a new airport. Inevitably, a cement factory is planned.

Although Ras al-Khaimah has no oil production as yet, a vast new cement plant capable of producing 500 000 t annually is being built. In addition, there is a small factory that produces aluminum housewares such as pots and pans.

The dominating concern of the sheikhs of the U.A.E. is to promote industrial development in their countries so that when their oil is all sold, they have a solid economic base to fall back upon. To this end, they are particularly interested in the desalination of seawater for drinking and industrial purposes and the establishment of oil refineries and petro-chemical works. They do not want to export jobs with the oil any more than they have to. This is the reason for a 1000-t-a-day nitrogenous fertilizer plant at Qatar powered by natural gas. This fertilizer will be sold abroad, as the output is too great for the U.A.E. to utilize by itself. Agriculture has no forseeable hope of expansion there, so to create wealth, more jobs must be found.

The tiny country of Sharjah has as yet not done very much with its new-found wealth but there are plans for roads, industry, and electricity. Education is not a concern. Since 1936 there has been compulsory education

Land use in Some Arab Countries:

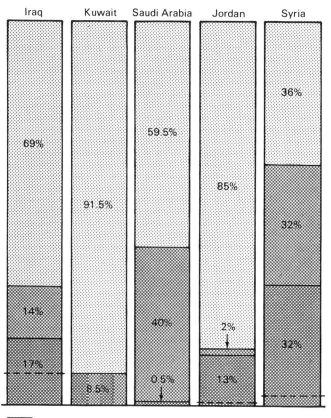

Iraq   Kuwait   Saudi Arabia   Jordan   Syria

Iraq: 69%, 14%, 17%

Kuwait: 91.5%, 8.5%

Saudi Arabia: 59.5%, 40%, 0.5%

Jordan: 85%, 2%, 13%

Syria: 36%, 32%, 32%

- ░ Desert
- ▒ Pasture Land and Forest
- ▓ Arable Land
- --- % Irrigated

through to the sixth grade, and in 1973 there were 100 university graduates and 1000 people who had gone through high school. Not bad for a country whose population in 1973 was estimated at 70 000. Sharjah has 80% of all the Arab graduates from university in the U.A.E.

Apart from oil, another precious liquid is found in Sharjah: water. There are abundant underground supplies trapped by sandstone and limestone. It is planned to build pipelines to the neighbouring states of the U.A.E. so that they will have a more secure supply of fresh water.

THE OTHER ARAB COUNTRIES

Away to the northwest, at the head of the Persian Gulf,

lies the country of Kuwait. Before the discovery of oil, the people lived by trade, fishing, and pearling. In fact, in 1912 Kuwait had the largest number of pearling boats of any state: 812, manned by 30 000 sailors.

Things had changed greatly by the early 1950s. Per capita national income was rising (it had been $35 in 1939) and industrial growth was coming about. Kuwait for a long time had the largest reserves of oil of any nation in the world. With a population of only about 837 000 in 1972 and a land area only one six-hundredth that of Canada, Kuwait seemed in an enviable position.

The Burgan oilfield is the largest single pool of oil found anywhere in the world. You will notice that it, like the other oil fields in Kuwait, lies very close to the sea, so that it is easy to exploit and export.

Just west and south of Kuwait lies the Neutral Zone. This is shared with Saudi Arabia after a treaty was drawn up in 1922. Oil was not exploited, however, until just after the Second World War. Like the other Arab Gulf

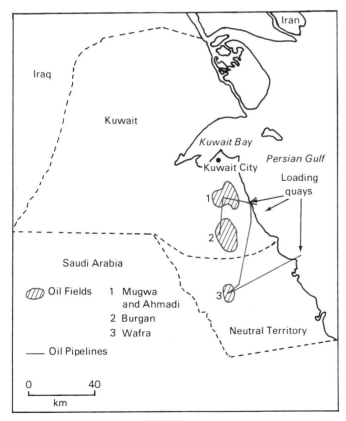

Figure 12-5. Kuwait's Oilfields

In 1972, 15 Iranians drowned while trying to swim over 14 km to shore after having been dropped at the edge of Kuwait's territorial waters. Forty-one others made it to shore through shark-infested seas, only to be arrested. Another time, an Arab from The Yemen was picked up suffering from exhaustion and put in hospital. He had been told that he could make up to $36 000 annually in Kuwait. Even though a more realistic figure was $2600 annually, it was a lot more than he could have earned at home. The big attraction for illegal immigrants is Kuwait's per capita income — about $5000 annually, the highest in the world.

states that have discovered oil, Kuwait is the target of much immigration, chiefly from other Arab countries. Many of the immigrants are in the country illegally. More than two-thirds of the labour force is non-Kuwaiti, although non-Kuwaitis make up only half the total population.

One of the greatest benefits coming from oil revenues for Kuwait is the assured supply of fresh water that sea-water desalination plants permit. Until the early 1950s, Kuwait suffered from a severe shortage of fresh water. For example, in 1947 17 500 l a day had to be brought in by ship from the Shatt al Arab, the mouth of the Euphrates and Tigris Rivers. Desalination plants were constructed in 1950, 1953, 1955, 1957, 1960, 1970 and 1971. Today these plants together can produce over 9 000 000 l of drinking water every day.

The industry that seems likely to benefit most from oil revenues is fishing. Shrimp and seafood, found in abundance, are exploited by three companies, each with its own packing plants. Ninety per cent of the shrimp goes to the U.S.A.

Socially, the Kuwaitis have benefitted enormously. Education is compulsory from 4 until 16 and is free. Kuwait University has several thousand students with a projected enrolment of 15 000 ultimately. There is a temporary but severe shortage of teachers in Kuwait. In 1969 there were 5104 of them, but only 68 were Kuwaitis. This position is bound to improve with time.

It is perhaps in the housing field that Kuwaitis take most pride. Low-income families are given free land for homes and interest-free loans to construct them. For those who can afford to buy, government-built houses have three main rooms, a kitchen, washroom, main sewage, electricity, and water supply for the shower stall as well as for drinking. (Bath-tubs would waste too much water.) In many respects, Kuwait has set the standards for other Arab countries to follow. It was Kuwait, for example, that first set up a fund to lend money for the development of other Arab nations. Nearly every Arab country in the world has borrowed money from this fund to finance development projects: irrigation, power stations, fertilizer plants, etc.

At the other end of the Arabian peninsula from Kuwait lies Muscat and Oman. Oil was first struck in 1964 and first exported in 1967. As yet, exploration is incomplete, but reserves seem ample for some decades. The revenues would go far to alleviate diseases such as trachoma, leprosy, malaria, and tuberculosis, as well as a variety of fevers. Water supply is problematical, and the water itself often impure. Sanitation is generally poor — often, in fact, a health hazard. There is no educational system as yet.

Attitudes are very interesting if not particularly unusual throughout the Arab world. The ruler is an absolute monarch. There is a low value placed on education, and women are excluded from public life. Smoking is prohibited on the streets and alcohol forbidden to Omanis. Muscat is probably the last city in the world where the inhabitants are locked inside the gates for the night. A curfew reigns and curfew-breakers are liable to arrest by the police. Change, if it has not already begun, is likely to be very severe in Muscat and Oman, adhering as it does to the traditional Arab values as spelled out in the Mohammedan *Koran.* Even slavery still survives in the interior of the country.

We turn now to the largest country in the Arab world, Saudi Arabia. Possessed of some of the largest oil reserves in the world, this country has immense influence and power. It has come a long way since it was the scene of battles against the Turks by the Arabs led by Lawrence of Arabia in the First World War.

More than the other Arab states, Saudi Arabia has encouraged the development of agriculture. Large new farms using artesian water from deep in the ground produce yields of wheat and other cereals every bit as high as Western countries. The influx of foreign oil-field workers has resulted in demands for all kinds of exotic foods such as roast beef and potatoes, curry, and many others.

Shell Photographic Service

Oil company housing for expatriate employees, Mina-al-Fahal, Muscat and Oman.

Shell Photographic Service

Mina-al-Fahal. The major oil terminal of Muscat and Oman.

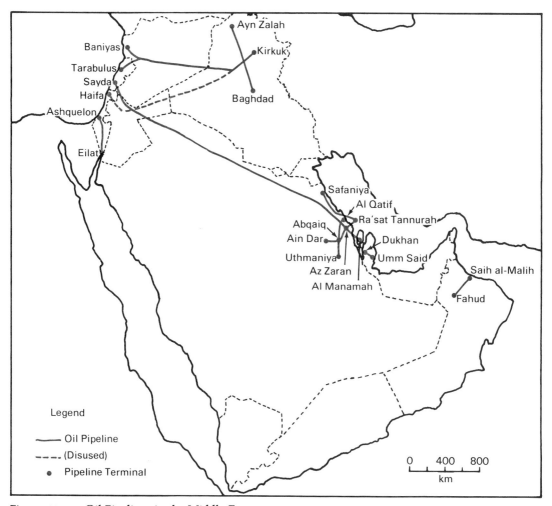

Figure 12-6.   Oil Pipelines in the Middle East

Fertilizer produced within Saudi Arabia has resulted in high crop yields being maintained.

Nomadism in the country has declined greatly, as many tribesmen opt for the advantages of life in cities. Still, many Bedouin prefer their existence as free men in the desert. Dependent on the winter rains for a flush of green in the desert valleys known as *wadis*, the nomadic life is a struggle. Pipelines, roads, and railways have made migrations more difficult than they used to be. It isn't easy to lead camels over a pipeline that is over 1 m in diameter, supported on piers perhaps another metre or so above the ground surface. International boundaries have hardened and are patrolled regularly.

As you can see from Figure 12-8, most of the major

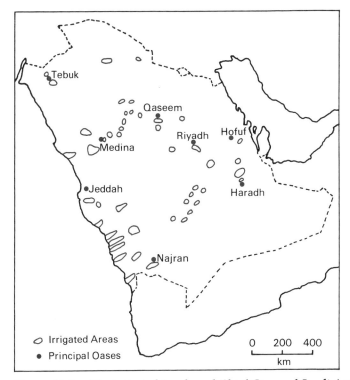

Figure 12-7. The Irrigated Lands and Chief Oases of Saudi Arabia

	Crude Petroleum		Natural Gas	
	Proven reserves	Production	Proven reserves	Production
Bahrain	74 000	3 761	23 000	314
Egypt	138 000	14 731	57 000	85
Iraq	4 420 000	83 775	566 000	870
Kuwait	10 370 000	146 786	1 119 000	4 041
Neutral Zone	1 901 000	28 970	195 000	210
Oman	649 000	14 689	62 000	3 613
Qatar	626 000	20 453	229 000	1 104
Saudi Arabia	18 737 000	223 412	1 918 000	no data
Syria	189 000	5 289	no data	—
United Arab Emirates	2 222 000	51 731	304 000	10
Iran	8 265 000	223 921	3 681 000	15 615
World	76 200 000	2 399 400	49 900 000	1 142 400

Figure 12-8. Proven Reserves and Production of Petroleum and Natural Gas in the Middle East, 1971.
Petroleum is measured in thousands of tonnes.
Natural gas is measured in millions of cubic metres.

Arab oil producers are Persian Gulf states. Their importance to the world cannot be exaggerated. They are members of the Oil Producing Exporting Countries, OPEC for
short. The actual members of OPEC are *Algeria, Libya,*
Iran, *Iraq, Kuwait, the U.A.E., Saudi Arabia,* Venezuela,
Indonesia, Nigeria, Ecuador, Trinidad, and Tobago. The
countries in italics are all Arab nations. Before we go any
farther, let's just see how much oil is controlled by these
OPEC countries.

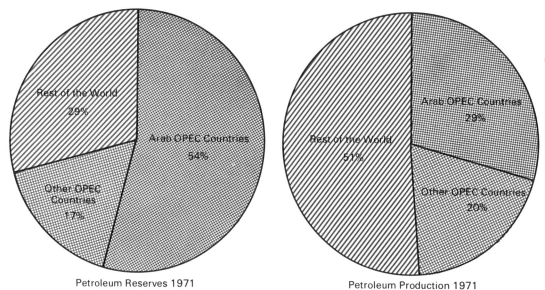

Petroleum Reserves 1971   Petroleum Production 1971

Figure 12-9.   World and Arab Oil Reserves

It is quite clear that there must be a limited supply of
oil in the world. A large part of the already discovered
known total proven reserves of conventional oil is actually present in the ground in the Arab countries who are
members of OPEC. When it comes to oil production, these
same Arab members of OPEC produce oil for export at
a rate less than that of the rest of the world. In other
words, Arab oil constitutes a large part of the world's
reserves and is being used up at a much slower rate than
the rest of the world is using up theirs. It seems it is
only possible for the world to continue using up oil resources at current rates *if the Arab countries agree.* We
are all aware from the press, radio, and television that the
Arabs are not prepared to carry on exporting oil blindly.
Why not? There are several reasons.

1 In 1971, world consumption of oil was 2 399 400 000 t.
Proven world reserves were 76 200 000 000 t. A little

Canada's proven conventional
oil reserves (excluding the tar
sands) in 1971 were sufficient
to last only 17 years at 1971's
rate of consumption. In 1974,
the period of time before exhaustion of proven reserves had
dropped to 7 years. However,
there are probable supplies in
the Arctic.

arithmetic shows that, at this rate of consumption, there is only enough oil to supply the world for 32 years. But the rate of oil consumption is increasing, so if no new reserves are found then the world's oil is likely to run out in a much shorter period than 32 years. In 32 years, a baby can grow up, marry, have children and see them off to school. In that space of time, the world's oil is going to be finished (according to the figures just quoted). How can you plan effectively if you know that your standard of living is likely to drop steeply after 32 years (probably sooner)? This is the situation that the Arab countries find themselves in. Even though probable reserves could multiply by 6 the 1971 proven oil reserves, the Arabs are worried. Even though a country like Kuwait has, according to Figure 12-8, enough oil to last 71 years at 1971's rate of production, not counting other reserves that might be found, the Kuwaitis are worried. They know their oil has to run out sometime. It might be when it is all used up, or it might be when alternative sources of energy are found, making their oil useless for that purpose. It is unlikely that significant oil deposits remain to be discovered around the Persian Gulf. If any new oil is likely to be discovered, then it is probably going to be on the continental shelves off countries such as Canada, the U.S.A., China, West Africa, and Europe. The Persian Gulf has already been combed pretty thoroughly. So the Arabs want to get as much money for their oil as they can now. It is pretty well their only resource. This explains why they and their fellow members of OPEC pushed up the price of crude oil from $2.59 a barrel to $11.65 in 1973 alone.

The reason why the Arabs did not attempt to raise the price of crude oil in earlier years is because they were not organized to do it. The Arab states were perpetually bickering among themselves. In addition, the major oil companies who exploited the oil reserves had the option of buying their oil elsewhere. But once OPEC got started in the mid-1960s, all the world's major oil-exporting nations belonged to it and they presented a united front to the international oil companies' tactics of playing off one country against another.

2 Rather than simply exporting oil in a crude state, and receiving in return a great deal of money, the Arab countries realized that they were exporting jobs. Here's the way it works. One oilfield worker can produce enough oil to keep several tanker crews employed, several oil refinery

employees employed, marketing teams busy, directors of companies busy, etc. The money used to pay these additional people represents *value added*. Why, reasoned the Arabs, do we not add value to the oil here in our own countries? That way, we would have more jobs than merely the ones for the oilfield workers. We would have oil refineries turning out fertilizer, pesticides, drugs, gasoline, heating oil (for export, obviously!) as well as a great many other things. The Arab nations would become great industrial centres. Their quality of life would improve. There would be less unemployment and the Arabs themselves would feel that they would be contributing to their own nations.

In fact, the whole trend of Arab governments recently has been to nationalize or buy out a controlling interest in the major international companies who operate there. This has turned the oil companies into contract buyers of crude oil. Iran, although a non-Arab country, is a member of OPEC and points the way most Arab oil-producers will follow. Iran has its own National Iranian Oil Company (NIOC). In addition there are 4 Iranian chemical and petro-chemical complexes. These draw their crude oil from NIOC and from foreign oil companies who are still doing business in the country but who are obliged to sell a portion of their output to NIOC at rates set by Iran. Finally, Iran is likely to invest in companies making oil products abroad, including in the U.S.A.

The Arab states, unlike Iran, do not have the industrialization that can absorb the vast oil revenues. So far they have not been able to do more than negotiate for control of the operations of oil companies in their own countries. In 1972, Iraq nationalized 55 foreign oil companies. Well before 1980 it is expected that Saudi Arabia, Kuwait, Iraq, Qatar, and Abu Dhabi will control over 50% of all the foreign companies operating in their lands. They already have 20% interests. Although they are paying for it, the billions of dollars are easily found from the oil revenues.

3   The last reason for the Arabs to put the brakes on selling oil abroad is - - - - - - . Can you guess what - - - - - - stands for? Six letters — the first is I, the last is l. Israel! But why is Israel never referred to officially by the Arabs? And how has it affected the oil supply from the Arab nations? Quite simply, Israel has had several wars with Egypt, Syria, and Jordan. It now occupies parts of all 3 of these countries. The Arabs have tried to have them re-

Shell Photographic Service

A geological field survey crew in Iran. What are they looking for?

If you travelled to an Arab country with an Israeli stamp in your passport in 1972, you could have been thrown into jail.

turned by force of arms and failed. They only lost more land when they tried. So Egypt, Syria, and Jordan asked their fellow-Arabs in the oil-rich countries to help them by cutting back the supply of oil to the Western countries of Europe and North America, as well as Japan. In this way, the Arabs hoped that these nations would put pressure on Israel to force the Israelis to give up the land they had occupied. Any country that did not openly back the Arab cause would lose its oil supply from the Middle Eastern Arab nations. That is what happened to the Netherlands. Eighty per cent of its oil supplies were cut off for a while in 1973. The rest of western Europe also relies for 80% of its oil from the Arab countries of the Middle East, Japan for 47% of its oil, and the U.S.A. for 10%, a figure which is expected to rise to 30% by 1980. The closure of the oil valve by the Arab countries, for whatever reason, would have severe consequences for the rest of the world.

Meanwhile, the Arabs grow rich. It is not easy to imagine such vast sums. In 1974 alone, Saudi Arabia earned $15 billion in oil revenues. Of that, only $3 billion could

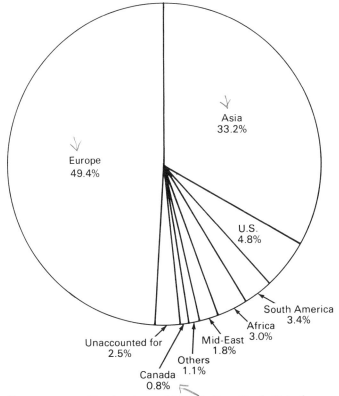

Figure 12-10. Destinations of Middle East Crude Petroleum, 1972

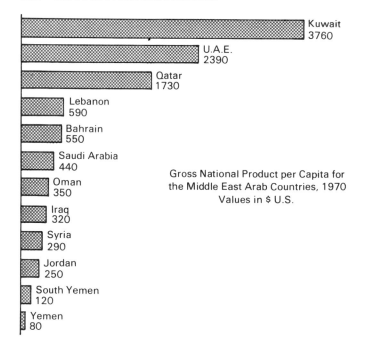

Gross National Product per Capita for
the Middle East Arab Countries, 1970
Values in $ U.S.

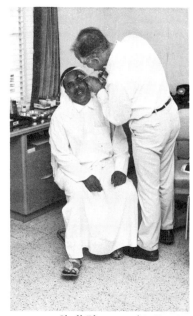

An oil company doctor and patient, Qatar.

be spent on internal development. It is estimated that by 1980 the country will have accumulated $100 billion in oil revenue that it will be unable to spend. Sharjah earned about $60 million in 1974. Abu Dhabi is expected to have an annual income of *at least* $4 billion by 1985. Annual oil revenues for Kuwait are about $10 billion.

Beside support of the affluent Arab nations for their poorer brethren's struggle with Israel, the countries around the Persian Gulf spend large sums of money on disputes among themselves. For example, in 1961 Kuwait became fully independent from Great Britain. Inside a week, Iraq (a socialist country) claimed the country. Only the intervention of British troops forced the Iraquis to abandon their claim temporarily.

In 1971, the British pulled out of the Persian Gulf area for good. As soon as they had gone, Iran stepped in and took over the islands of Abu Musa and Greater and Minor Tunb. (See Figure 12-3.) In the face of this action, Iraq broke off diplomatic relations with Iran. Part of the problem seems to be that Iran and most of the Arab states are ruled by sheikhs or kings. They are hereditary rulers and very different from the socialist government of Iraq, which receives large shipments of arms from the Soviet Union, as do Egypt and Syria. At any rate, in 1973 Kuwait bought $500 million worth of U.S. tanks, missiles,

In 1974, the U.S.A. decided to open up embassies in Bahrain, Qatar, Muscat and Oman, and the U.A.E. In addition, they planned to help Saudi Arabia develop large-scale industries. Among other things, a giant steel mill is planned at Jubayl on the Persian Gulf coast. Petromin, the Saudi state oil and minerals firm, two Japanese steel companies, and a Dutch pipeline firm are to be in on the venture, too. All told, it will be worth $5 billion.

and military construction assistance. This was just after the U.S. had sold $2.5 billion worth of arms to Iran, and Britain had disposed of $625 million worth of airplanes to Saudi Arabia, which also bought $1 billion worth of U.S. Phantom aircraft.

Another factor that contributes to internal tension within Arab oil-producing nations is that most of these nations rely on outsiders to help plan and develop their economies while their young people (at least, some of them) are away in other Arab countries gaining an education. Usually they go to Egypt or Syria, where they often come into contact with *ideologies* that run counter to those at home. Add to this problem the Dhofar Liberation Front, which operates in western Oman and aims to stamp out hereditary rulers in the U.A.E. as well as Muscat and Oman. Include the Palestinian Arabs, who demand the return of their homeland from Israel. You can see that here is an explosive enough mixture, quite apart from the presence and implications of the large deposits of petroleum. The Middle East is in the throes of rapid change, all right. Just what the future holds in store is anyone's guess.

QUESTIONS AND EXERCISES

1. Construct climate graphs to show the data relating to Aden and Baghdad. Compare these data with the climate of your own area. Write down any conclusions that you come to.
2. Write up a project of about 2 pages on the way in which the Bedouin Arabs survived in the desert. You will have to do your own research on this topic, and probably will have to use your own library.
3. The density of population in Canada is about 2.4 people per square kilometre. How do the population densities of the countries in Figure 12-4 compare? Show your results in the form of a graph or series of graphs.
4. Calculate how long 1971's reserves of oil and natural gas are likely to last in each of the countries listed in Figure 12-8, assuming the rate of consumption stays at 1971's level. Is that a fair assumption to make?
5. Change the percentage bar graph of Figure 12-10 into a map with arrows whose widths are proportional to the percentages given. The arrows should also point in the right direction.
6. Do you think the Arabs' tightening of control over oil resources found in their lands is a reasonable or unreasonable thing to do? Give reasons for your answer.

13. ISRAEL

SUPPORT OF LIFE

Sitting on the bumping tractor, Isaac Solomons hummed as he squinted along the furrows in the heat of the day. "I must keep them straight," he thought. "If I don't, the management committee will never give me this job again." He bit his lip. He knew he should not think like that, living on a kibbutz. "From each according to his ability, to each according to his need." It sounded just fine. But there were still jobs a person preferred, eh? After all, he was only human!

He came to the turnaround at the end of the field. He had had strict instructions. "Whatever you do, do *not* plough past the markers. We haven't checked there yet and you must wait until the sappers have given us the go-ahead. We're pretty sure it's okay, but you never know. Those Syrians are very clever when it comes to laying mines."

Here it was. A short, red post, one of a line. Plenty of room to take a tractor through between them, though. And we do need the extra land. I'll give it a try. After all, other people are being killed every day. They risk their lives and I'm not allowed to?

The tractor swung between the red posts on its way up the field. Sol Weizmann snorted. "What an idiot! He has been told not to do that, the numbskull!" Angrily he started down from the low rise where the kibbutz buildings lay. Muttering oaths as he picked up speed in the jeep, he swung around a turn and saw Isaac on the return trip up the field. "I'll wait till he gets back level with here," thought Sol. "Then I'll tell him to obey orders!"

Isaac was blissfully ignorant of the commotion he had caused. He didn't notice the jeep with the dun, desert camouflage, nor the angrily erect figure beside it. Here we are again! He turned the tractor through 180°. The challenge of the line of red posts swung into view, and once more he was intent upon keeping a straight furrow. Then the earth beneath him opened up in an orange rose. A deafening roar, and he knew no more.

Sol raced across the field, but he knew he was too late. Isaac had been blown to pieces, along with the tractor, by a mine that had been missed in the search the day

In October 1972, an Egyptian security officer was severely injured when he opened a letter addressed to the leader of the Palestine Liberation Organization, a movement of Palestinian Arabs who lost their homes on the West Bank of the Jordan either in the 1948 or 1967 wars. The security officer was suspicious of the letter, which was addressed to the PLO in Cairo.

before, or, more likely, that had been relaid in the night. He halted by a smoking, reeking crater. Nothing, nothing left at all. He lifted his eyes to the north east to a line of high hills. They looked lifeless but he knew better. Somewhere up there, he thought grimly, is the man who planted this mine and he's watching me right now, just as he watched Isaac die.

"Have this entire field swept for mines again," he ordered a shocked older man who stood panting beside him. "And have the remains taken away as soon as possible. We must never give up ploughing and cultivating this land. Otherwise they've won. And Isaac died in vain."

The scene above never did actually happen but it could well have done. The Israelis were, and still are, at odds with the Syrians and all the other Arab nations too. But it is vitally important that all the land possible be ploughed and planted. Israel is a small country, and some of the farm workers simply have to continue work in full view of their enemies. Every now and then, one of them may die as a result of enemy action. Why? Why all the hatred?

The answer to that question is rooted in history. It is possible to fill many volumes with the history of the tiny area known as Israel today, but by many other names in the past. Let us pick up the thread in the year 70 A.D. In that year, the Romans exiled many Jews after a revolt that was unsuccessful, and Israel was ruled as a Roman province. With the decline of the Roman Empire, the land came under the successive governments of the Arabs, the Turks, the Crusaders, the Mamluks, the Turks again, and Britain. Then a United Nations resolution in 1947 approved that the State of Israel be created as a homeland for the Jews.

This decision was most unpopular with the Arabs.

Masada brings a surge of pride to all Jews. It is a rocky plateau that was fortified by King Herod. Occupied by the Jews in the great revolt of 70 AD, it finally succumbed to the Roman legions in 73 AD, but all the defenders killed their families, then drew lots to kill one another. The last man killed himself. It was done to deny the Romans the chance of defeating them. The site of Masada was excavated by *Yigael Yadin*, the famed Israeli archaeologist. His book, *Masada*, is well illustrated and *worth locating in the library.*

The Rome representative of the PLO was killed in ambush in Rome in October 1972.

In November 1972, the Israelis launched an air raid on PLO bases in Syria in retaliation for the freeing of three Arab guerrillas by West Germany. They had participated in the Munich Olympic massacre of 1972, when several Israeli athletes were killed. The PLO guerrillas were freed after some of their comrades hijacked a Lufthansa airplane and demanded their release.

Israel Government Tourist Office

Herod's palace at Masada. Masada is a great isolated plateau in the desert which was fortified by King Herod. It was defended to the death by insurgent Jews who demanded freedom and independence from the Roman Empire. Since then, the ruins have been uninhabited. Nowadays, they attract many tourists.

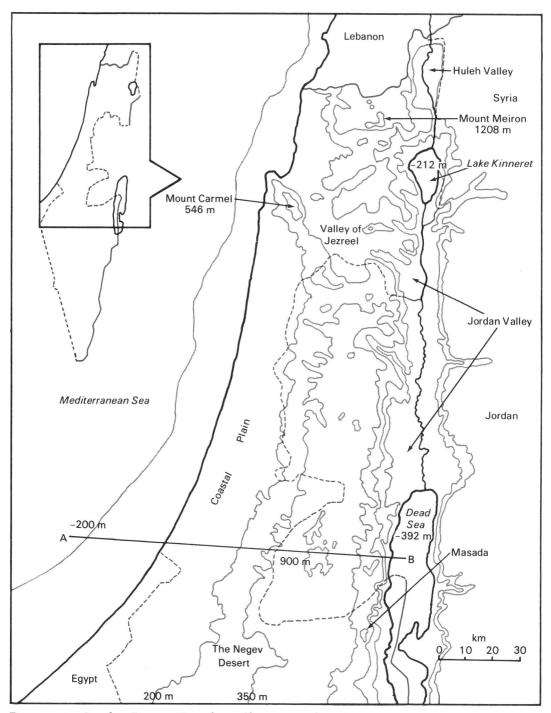

Figure 13-1. Israel — Pre-1967 Boundaries Shown as Broken Lines

Although they had agreed in 1919, at the end of the First World War, to give the land of Palestine (west of the Jordan) over as a homeland for the Jews, they became unsettled at the waves of immigrant Jews who arrived in Palestine, most escaping from the Nazis in Europe. Palestine had been handed over to the British in 1922 by the League of Nations. It had been part of the Turkish Ottoman Empire until the Turks were defeated. Under the British, Arab and Jew lived uneasily together with occasional flareups leaving dead and wounded on either side. In 1948, the year after Israel was created, the regular armies of the Arab countries invaded the country. Israel held out and achieved an armistice. It was signed in 1949. Peace has never been declared, and there have been three other wars in the meantime: in 1956, 1967, and 1973.

The map of Israel has undergone some drastic changes since 1948. It is rather hazy as to what constituted Israel in 1948, but the British had divided the area called Palestine in two. The land to the west of the Jordan River was known as Palestine, and the land to the east of the Jordan was given over to the Arabs as the *Emirate* of Transjordan that would later be known as Jordan. And then . . . but it's rather confusing to try to say it in words. Let's use the language of geography — maps.

After the 1967 war, Israel's area increased from 20 700 km^2 to nearly 90 000 km^2.

First, let's make it perfectly clear which Israel we are referring to when we talk about Israel. Since 1948, the longest lasting boundary has been the one drawn at the 1949 armistice. Any data that are used in this chapter are for that particular area of land unless stated otherwise.

POPULATION

In 1948 there were less than a million Israelis. This small figure quickly increased as immigrant Jews arrived from all over the world, but chiefly from war-torn Europe and the Arab countries. In fact, in the first few years there were more immigrants than births in Israel. In 1950 the *Law of Return* guaranteed the right of Jews all over the world to come to live in Israel if they so desired.

Because so many Jews came from so many different places, there were some startling changes in the national origins of the population. In 1948, 35.4% of the people had been born in Israel, 9.8% had been born in Africa or Asia, and 54.8% had been born in Europe or America.

Not all the inhabitants of Israel are Jews, although 85% are. In addition, 12% are Moslems, 2% are Christians, and 1% are Druzes, a Moslem splinter group.

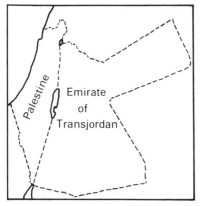

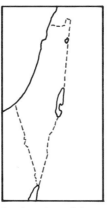

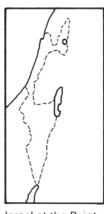

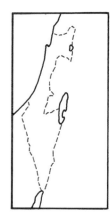

The British Mandate of Palestine –
Partition in 1922

Declaration of the
Establishment of
Israel, May 14, 1948

Israel at the Point
of the Farthest Arab
Advances, 1948

Israel at the
Armistice, 1949

There has been one more war since 1967. It was fought in 1973 but
there was little change in the boundaries except for some
gains along the Golan Heights and along the Suez Canal, which have
since been returned as part of a peace settlement. Today the
borders are substantially those shown on the left though, of
course, they could change at any time.

Israel at the Armistice, 1967.
Annexed Territories are Underlined

Figure 13-2. **The Changing Boundaries of Israel**

By 1972, the percentages had become 47.2, 25.7, and 27.1,
respectively.

A big problem involved with accepting so many immigrants was that not only was there a desperate shortage

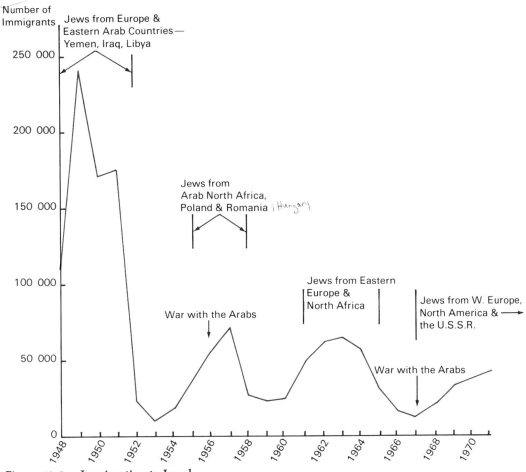

Figure 13-3.   Immigration to Israel

	Israel	Canada
1948	650 000	12 823 000
1956	1 629 500	16 081 000
1961	2 185 000	18 269 000
1966	2 629 000	20 050 000
1971	2 972 000	21 786 000

Figure 13-4.   The Growth of Israel's Population Compared with That of Canada

of housing, but many of the immigrants were from un-developed countries. This resulted in a cultural gap in Israel. It has been narrowed to a great extent by ethnic intermarriage, education programs, and almost compul-sory teaching of Hebrew, the national language. There are still some tensions, however, but doubtless these will ease

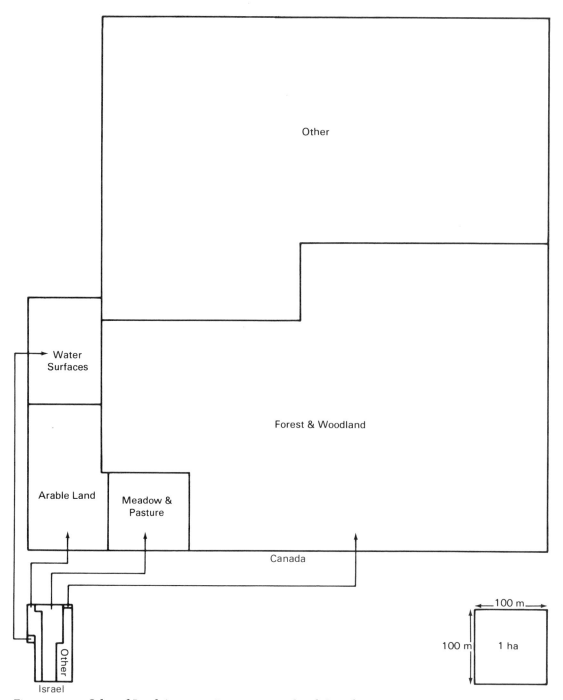

Figure 13-5.   Selected Land Areas per Person in Israel and Canada

Despite all efforts, there is a growing gap between rich and poor in Israel. Thousands of Israelis have no telephones, for example, and live in the slums of Southern Tel Aviv and Jaffa. Generally, the people who live in poverty in Israel are handicapped by their cultural background and lack of education, having immigrated from poor countries. By way of contrast, a European model car with a stick shift costs $5000, and the prices of homes can range up to $100 000, yet many people seem able to afford them.

with time. As the country is so small, its inhabitants will simply learn to live with one another. It would be different if there were vast areas where the Jews of different national origins could separate into groups. But there are no vast areas in Israel. You can drive across the country in 90 min from east to west, and in 5 h from north to south.

THE LAND

Obviously, there really is not very much land in Israel. Thus what little there is has to be used very wisely. In this Israel has performed well. The most important problem that faced Israel was not merely the small amount of land that could be used to grow crops and raise animals. It was the fact that Israel is mostly desert, or situated close to a desert. As a matter of fact, Israel is 60% desert, but only 6% of its people live there. It is vital that Israel try to establish more settlement in the Negev, because it represents the only remaining available land for cultivation using existing rainfall and irrigation.

Israel is already using over 90% of its water resources, so that further settlement of the Negev will have to wait until there is large-scale desalination of seawater. There is already a plant at Eilat on the Red Sea coast and another one is planned there, as well as an even larger one at Ashdod on the Mediterranean Sea coast. The construction agreements were signed with the U.S. in 1972. A possibil-

In 1971 and 1972, over 44 000 Soviet Jews left for Israel, making up over 60% of Israel's immigration in that period. The Soviets were reluctant to let them go for a variety of reasons, one probably being that since the U.S.S.R. backed the Arabs, they could not very well allow Israel's population to increase. So they imposed an *exit tax*. Soviet Jews who wanted to go to Israel had to pay back the cost of their education. The amount could have been as much as $25 000.

There is a housing shortage in Israel. In 1972 there were reported to be 51 000 empty apartments in the country. Yet many of the Soviet Jews who arrived refused to go to them, saying that they were too small or poorly located. This created resentment among native Israelis, who were also unhappy to learn that the newcomers were given tax concessions.

AVERAGE NUMBER OF RAINY DAYS

	J	F	M	A	M	J	J	A	S	O	N	D
Jerusalem	12	9	8	3	2	0	0	0	0	2	7	9
Tel Aviv	14	12	8	4	1	0	0	0	1	3	8	12
Haifa	15	12	9	5	1	0	0	0	1	4	8	12
Tiberias	12	10	5	3	1	0	0	0	0	1	5	8
Eilat	1	1	2	1	0	0	0	0	0	0	1	1

A severe drought hit Israel in 1973. Water reserves dropped to 6% of normal. Lake Kinneret dropped 1.5 m. Heavy winter rains in 1973-1974 saved the situation.

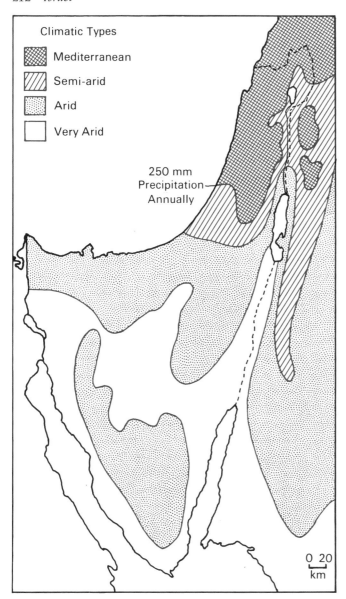

Climatic Types

- Mediterranean
- Semi-arid
- Arid
- Very Arid

250 mm Precipitation Annually

0 20
km

Figure 13-6.   Climates of Israel

Israel Government Tourist Office

In the depths of the Negev, the great desert that covers all of Southern Israel.

ity, but only a possibility at the moment, in the present mood of tension throughout the Middle East, is the use of nuclear power stations to purify or desalinate seawater, using their waste heat after generating electricity.

The settlement of the Negev illustrates the basic Israeli settlement types that were, and still are, used to pioneer and develop the land.

The *kibbutz* was evolved as the best way to tackle problems that an individual could not surmount on his or her

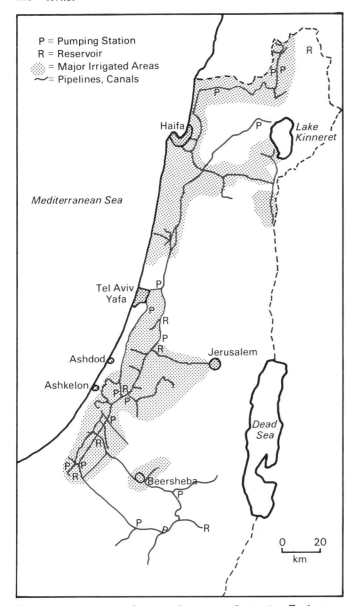

Figure 13-7.   National Water Carrier, an Irrigation Project

own. In 1897 an international Jewish movement called
Zionism attracted many volunteers for a return to Israel.
There was no question but that the land to be developed
would have to be bought from the Arabs who owned it,
and permission obtained from the Turks, who at this time
ruled Palestine as part of the Ottoman Empire. In 1909
the first kibbutz was established at Degania at the southern end of Lake Kinneret, with only 10 men and 2 women

As the night air cools, the ground surface cools, too. The loose rocks have a large surface area in relation to their volume, and they cool very fast. The air is chilled when it comes into contact with them, and water vapour is precipitated as dew, which runs downhill under gravity to collect around the base of the tree. There it sinks into the ground, where it penetrates to the roots before the hot sun of the daytime can evaporate it.

Loose rocks

A natural system of irrigation developed by the farmers of the Negev.

The Jordan River flows into Israel's Sea of Galilee and then southward to the Dead Sea. It is prominent in early Christian history.

to work it. Today there are over 230 *kibbutzim*, worked by nearly 100 000 *kibbutzniks*.

The ideals of the kibbutz are straightforward. All property except for personal effects is held in common, and money is never used, even in the dining halls and shops. These are communal, too, and all members of the kibbutz have to take their turns waiting at table and washing up. In fact the management committee, elected by all members, draws up a rotating schedule of work activities. Anybody who works outside the kibbutz pays his or her salary into the kibbutz treasury. Children are raised in age groups in dormitories. Those too young for school are tended during the day by the woman in charge of the baby house. Late afternoon is the time when parents and children mix together, and before sunset, the babies are returned to the baby house, and the older children are returned to their own quarters. If someone needs a new pair of pants, a pack of cigarettes, or anything else, then they are permitted to draw up to their limit from the store.

The *moshavim* are rather different from the kibbutzim. Each family has its own land but buys and sells co-operatively, some farm equipment being owned by the village as a whole. A general assembly approves or refuses the transfer of farms and new members. There are some 350 moshavim in Israel with over 120 000 members.

The *moshav shitufi* is identical to the kibbutz except

During the Arab-Israeli war of 1973, women and children were left to run the farms. Fifty per cent of all workers were called up, yet production was 65% of normal. Even so, to pay for the war, the average Israeli's income taxes rose by about 10%.

that each family has its own household and looks after its own children. There are less than 30 moshav shitufi, with nearly 6000 members.

Altogether only 8% of Israel's people live on communal farms, or in collective or co-operative villages, but they contribute about a quarter of Israel's GNP. In addition to their agricultural functions, some of them have added industries and tourism to their list of wealth-generating activities.

The inset map on Figure 13-8 shows another type of settlement, the *nahal*. Literally this means "fighting pioneer youth." The young men and women who are members undergo 6 months of intensive military training, then transfer to the countryside for a year where they work land that is too exposed to enemy bullets and shells for civilians. At the end of that time, they hand over to another group. As you can see, the nahals are all in frontier, or what were once frontier, locations. They were developed in response to a shortage of volunteers for the regular kibbutzim. Perhaps the coming to Israel of a comfortable, urban way of life made many people unprepared to face the rigours of pioneering on the frontier. The average number of people on the kibbutzim, as you can see in Figure 13-8, is only 200. The moshavim, situated as they are on more favourable land, have some 300 families each. Originally the moshavim gave each family 4 ha of irrigated land for vegetables and a piece of non-irrigated land for cereal crops. Later on, the emphasis changed to industrial crops of cotton, sugar beets, and peanuts. In addition, there is dairy farming. The whole aim of Israel's agricultural effort is to make the country as self-sufficient as possible, and in this it has largely succeeded. All the required fruits and vegetables are grown, as well as half the required amount of wheat. The land area under the plough has trebled since 1948, but the area irrigated has increased by more than 6 times. Surplus fruit, eggs, vegetables, poultry, milk, and dairy produce is exported. The major export of agricultural produce is citrus fruit, largely in the form of juice.

THE CITIES

Although so far we have discussed only the countryside and villages, Israel is, in fact, a highly urbanized society. More people live in cities in Israel than in most other lands — more, that is, as a percentage of the population.

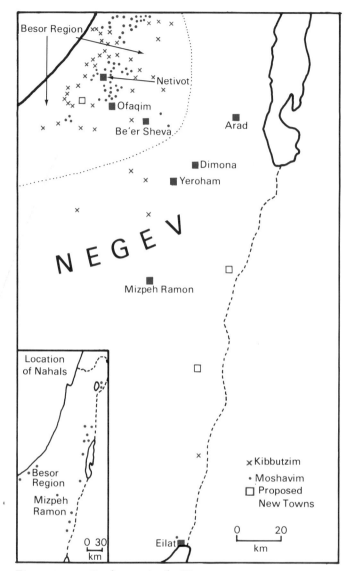

Figure 13-8. Settlement in the Negev

Israel Government Tourist Office

The relationship of Tel Aviv and Jaffa is clearly evident as is the contrast in styles between young brash Tel Aviv and ancient Jaffa. The small spire of the Jaffa Mina-rette stands out in sharp contrast to the square contemporary Sha-lom Tower in Tel Aviv. Jaffa is one of the oldest cities in the world. Tel Aviv is one of the youngest.

The figure for Israel is 82.5% and it is still rising. The reason for this is not hard to find. If you look at Figure 13-10, you will see that most people work in industry of some form or another.

Where there is manufacturing industry, there are towns and cities that serve as convenient focal points for labour and production. After all, you wouldn't set up a factory in the middle of uninhabited desert. Once the towns and cities are established, service industries come into being — filling people's needs for banking, insurance, food serv-

Pollution is worrying Israel. Lake Kinneret is threatened with sewage and fertilizer. Tel Aviv smog is proportionally worse than that of Los Angeles. Every river in Israel is a health hazard. Car exhaust fumes are everywhere. The coral reefs of Eilat, considered by some the best and most scenic in the world, are in danger from oil spills.

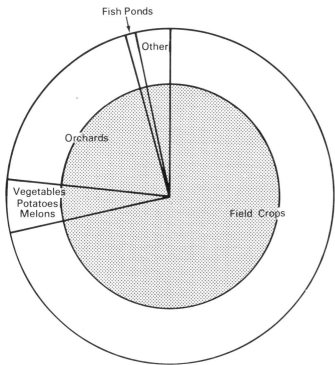

Figure 13-9.   The Growth and Character of Israel's
              Agriculture, 1948-70

Crop	Production	
	1948	1970
Citrus Fruit	272 200	1 513 500 t
Wheat	21 200	199 500 t
Cotton Lint	none	36 700 t
Tobacco	600	1 100 t
Vegetables, Potatoes, Melons	11 900	765 300 t
Poultry Meat	5 040	124 200 t
Cattle Meat	2 010	36 500 t
Fish	3 500	26 100 t
Milk (kilolitres)	85 950	497 500 t
Eggs (thousands)	242 500	1 423 200 t

ice (restaurants), and so on. In service industries, things are not actually produced, but services, which are worth money, are performed. The stamp of a developed economy is to have relatively few people in farming (because it is efficient), and more people in manufacturing industries, making consumer goods such as cars, televisions, and radios. Finally, there should be a lot of employment in

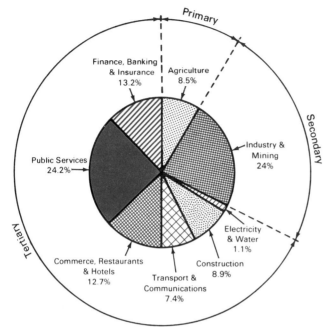

Figure 13-10.   Employment in Israel by Economic Group, 1971

industries that service people. Those three types of employment are actually called *primary, secondary and tertiary economic activity.* Figure 13-10 should make it clear to you.

It's logical, then, that the towns and cities of Israel have grown very rapidly. In fact, new towns and cities are springing up to accommodate the increase in population.

As Figure 13-11 shows, most of the new town and city growth took place in the wetter northern half of the country or along the coastline of the Mediterranean. Some

Israel Government Tourist Office

Haifa, one of Israel's newest cities, is seen here from Mt. Carmel. How can you tell this is an oil terminal?

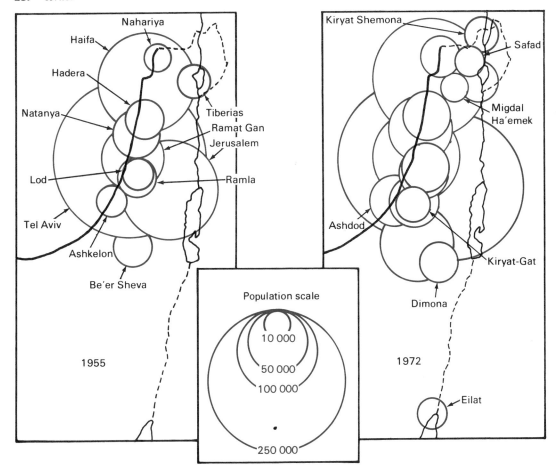

Figure 13-11.   City Growth in Israel

In 'Akko it is demonstrated that Arabs and Jews can live peaceably together. The city has 36 000 people, 9500 of them Arabs. They have 3 representatives on the city council and a vice-mayor. During the 1973 war, hundreds of the city's Arabs donated blood, money, and time to help the Israeli war effort. Of course, there was sabotage and some espionage rings were uncovered from time to time, but because the Jews and Arabs live together in 'Akko, unlike most other Israeli cities, the resentment and distrust soon disappeared, with constant daily contact between the races acting as a balm. "I wouldn't say all the Arabs love us, but living together blunted the edge of the old hatred," said one member of the city council.

of the towns have specialist functions — for example Eilat and Ashkelon, which are the termini of an oil pipe-line, by-passing the Suez Canal which was closed to

Israeli ships by Egypt in 1956, and blocked to all traffic in 1967. (It is scheduled to reopen soon, though.) By far the greatest impetus to Israeli urban development has been the need to accommodate large numbers of immigrants in a planned way, so that they will have homes and jobs and feel productive, valued members of society as soon as possible.

NATURAL RESOURCES

There are very few natural resources in Israel. Copper is found at Timna near Eilat, but production is not very great. Phosphates for fertilizer are found at Oron near Yeroham. Production runs at about 1 000 000 t annually. Potash and other minerals are extracted by evaporation from the very saline waters of the Dead Sea. There are small amounts of oil and natural gas, chiefly at Arad near the Dead Sea, where a large new petro-chemical and chemical plant is being worked up to full production. The Heletz district also has some oil.

The administered territory of Sinai might have some hidden riches though. There is a small oilfield at Belayim that produces most of Israel's needs. Oil moves around the coast of the Red Sea by tanker at Eilat. Most of it is exported, though, because it is heavy and sulphurous and the Israelis prefer to sell it to Italy and import oil from Iran. The oil is worth at least $35 million a month to Israel which is one reason why Israel does not want to hand Sinai back to the Egyptians.

Just after the 1967 war, the Israelis surveyed Sinai for minerals, but the report has not been released. Interim reports show that there is some tungsten, copper and feldspar, bauxite and water. The presence of water is merely a rumour, but it has not been denied. There might be as much as 200 billion cubic metres of underground water with a low salt content. Even if only 5% of that is accessible from Israel proper, it would be sufficient to support 2 million people. But Israel holds all Sinai. Again, if true, this is another reason why Israel would be most reluctant to give up the peninsula. Yet this is what Egypt demands.

INDUSTRY

Considering the nature of the country and the lack of raw materials, as well as the small but generally well-educated and technically-minded population, it is natural for Israel to concentrate on small products with relatively little raw material in them but with a high proportion of value added by skilled labour. Nearly 10% of total industrial production is taken up by *electric* and *electronic equipment*. *Textiles and wearing apparel* are also very important. The large British firm of Marks and Spencer has a great deal of its clothing made in Israel, which is then exported to stores in Britain and Canada. *Food and food products* are also very important industries. The cutting and polishing of *diamonds* has grown remarkably. Sales abroad rose 50% between 1971 and 1972 alone. Israel has to pay a lot for the raw material, of course, but even so the value added to the diamonds by cutting and polishing them makes the diamond industry equal in importance to that of food.

The structure and efficiency of the economy is such that per capita Gross National Product in Israel was $2200 in 1972. This compared well with the Netherlands ($2000), West Germany ($4000), and other developed nations. It was far ahead of Syria ($290), the Lebanon ($590), Egypt ($210), and Jordan ($250).

Computers: 1962 —   6
             1971 —   257

Electricity annual consumption:
             1948 —   70 000 000 kW.h
             1970 —   1 945 000 000

Motor vehicles in use:
             1949 —   15 800
             1971 —   296 299

Phosphates production:
             1954 —   58 000 t
             1969 —   1 026 000 t

Potash production:
             1947 —   103 000 t
             1970 —   650 000 t

Exports: 1950 — $ 35 000 000
             1971 — $957 000 000

Petroleum products consumption:
             1950 —   650 000 t
             1969 —   4 151 000 t

Figure 13-12.   The Israeli Economy: Some Indicators

FOREIGN TRADE

In order for the economy to grow, Israel must expand its

	1960	1971
Passenger ships	20 388	2 023
Cargo ships	291 982	1 284 454
Refrigerating ships	—	89 638
Tankers	61 809	2 100 679

Figure 13-13.   Israel's Merchant Fleet in Tonnes (Deadweight)

export markets. The home market is too small to allow for much expansion. A by-product of this trend to export has been an expansion in the Israeli merchant shipping fleet. It has now reached the point where there is a shortage of manpower. Roughly one-third of all the freight carried is taken between foreign ports, making Israel's merchant fleet an international carrier.

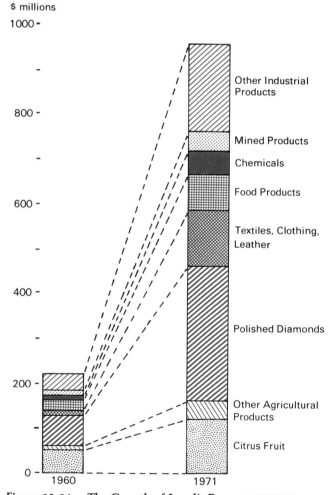

Figure 13-14.   The Growth of Israel's Exports, 1960-71

The most important goal of Israeli trade is to balance its books — that is, to export at least as much as it imports. It consistently has not done so, but it has been the receiver of capital from other countries, so that on balance the accounts often come out in Israel's favour.

Mention must be made here of *remittances* from other countries — that is, gifts of money from Jews outside Israel to allow the country to survive. The money is often raised by fund-raising expeditions to North America. In the late 1940s, the money was used to buy arms.

As you can see from Figure 13-14, there are more industrial products exported than agricultural products. This is only to be expected, for Israel has a modern, technologically oriented type of economy. You can also see, then, that Israel would import few finished goods for direct consumption because they would have to be paid for out of exports. Israel exports mainly finished goods. It would make no sense for the country to buy finished goods such as transistor radios and clothes when it can obviously make them for itself. In 1971, Israel imported $179 million of finished goods but $1113 million of raw materials. *Investment goods* (ships, aircraft, machinery —

	Visible Imports	Visible Exports
EUROPE		
United Kingdom	276	98
West Germany	235	91
Netherlands	79	58
Belgium	74	44
France	85	42
Switzerland	62	43
Italy	85	23
Sweden	29	13
Romania	27	11
ASIA		
Japan	58	48
Hong Kong	2	45
Iran	3	33
AFRICA	25	47
AMERICA		
U.S.A.	430	186
Canada	21	16
South America	30	13
OCEANIA	4	7

Figure 13-15. Imports to and Exports from Israel by Main Countries, 1971 ($ millions)

anything that can be used to make other products) totalled another $1809 million in addition.

Perhaps you now have a better idea of the way in which the people of Israel support themselves. In case you wanted to see for yourself, you could go there as a tourist. El Al, Israel's international air carrier, carried over half a million people to Lod airport in 1972, from nearly every part of the world. You could stay in any of the more than 300 recommended and government approved hotels and other accommodations. Ancient sites of historical interest, breath-taking scenery, natural wonders such as the Dead Sea, skin-diving in the Red Sea, sailing on Lake Kinneret. There is a lot to see and do. No wonder a million people visited Israel in 1974.

QUESTIONS AND EXERCISES

1. Draw a cross-section along the line A-B in Figure 13-1. Label everything you can on your cross-section and find out what you can about the Dead Sea and how it differs from Lake Kinneret. In addition, find out what a rift valley is and label one on your cross-section.
2. Construct graphs to show the number of days with rain at Jerusalem, Tel Aviv, Haifa, Tiberias, and Eilat. Total the number of rainy days for each place and mark the total on an outline map of Israel in the correct place. See if you can divide up Israel by drawing lines at intervals of 5 days.
3. Construct divided circles to show the national origins of Israel's population in 1948 and in 1972.
4. Draw a graph to show the growth of Israel's population and Canada's population, using the figures shown in Figure 13-4. Then convert the actual figures for each country into *index numbers* and plot them on a graph. How do you explain the difference between the two graphs?

    Formula to calculate the index numbers.   1948 = 100

$$\frac{\text{Index number}}{\text{value for 1948}})\text{value for any other year} \times 100$$

5. Find out what you can about the following expressions, names, etc.: *Yishuv, Theodor Herzl, Knesset, Sephardim* and *Ashkenazim, Sabra, Olim, Jew.*
6. Devise a way of showing the information in Figure 13-9 graphically.
7. Devise a way of showing the information in Figure 13-15 in the form of a map. Distinguish between *visible* and *invisible* trade.

14. MIGRATIONS

SPATIAL INTERACTION

Back in the mists of antiquity, there were a series of vast migrations of people. From the heartland of Asia, great waves of migrants moved to populate Europe and the Americas. The waves were periodic, separated by centuries, but over the millennia enough people moved to cause the birth of population clusters in Europe and the Americas. The eastern and southern fringes of Asia itself were also populated by these pulsing migration waves.

But these migrations of long ago were as nothing compared to the migrations that have taken place in recent years in Eurasia. Nowadays, millions of people are involved, not just thousands; years, even months, are involved, not centuries. The whole pace of migrations has picked up, so that there are now more migrants than ever before, moving more quickly than ever before.

INTERNATIONAL MIGRATIONS

There are three different aspects of migration in Eurasia. First, migrants move from one part of Eurasia to another, across international frontiers, such as East Germans escaping across the Berlin Wall. Second, migrants move from Eurasia to another region of the world altogether, such as Chinese moving from Hong Kong to Toronto. Third, migrants come into some parts of Eurasia from elsewhere, such as Algerians coming to work in the European Common Market. Let's deal with them in turn.

Marilyn Olsen

The Berlin Wall.

1 Migrations between different countries in Eurasia

There have been 4 or 5 major migrations of this type in recent years. Some of them are still continuing. The longest lasting (since World War II) has been the movement of people from the communist lands of eastern Europe to various countries in the so-called West. It is estimated that about 20 000 000 people have been involved in this migration, making it one of the most important in Eurasia. The migration itself has not been simple or uniform. In fact, there have been several different aspects to it, and several different destinations. The most publicized aspect has been the persistent stream of people from East Germany across the Berlin border, causing the East German authorities in 1961 actually to build a wall to try to contain the escapees. But the escapes continued, often bringing on a barrage of rifle and machine-gun fire as people tried to scale the wall. People have also swum rivers to get to the West, hi-jacked aeroplanes, stolen cars and buses and rammed through road-blocks, tunnelled under obstacles; they've used all their ingenuity to get out. Most of them made it safely, but many were killed. If your teacher can get the film, *The Wall*, it's well worth watching.

Another publicized aspect of this same migration (from the communist lands to the West) has been the attempts by the Jews in the U.S.S.R. to leave and go to Israel. The U.S.S.R. has not actually stopped the Jews from leaving, but it has done everything it can (which is a lot) to make it hard for them to leave. Nevertheless the Soviets admit that about 95% of all the Jewish applicants actually gain permission to leave eventually. The Jews who succeed in leaving the U.S.S.R. all pass through transit camps in Vienna. Curiously, they pass Jews going the other way back to the U.S.S.R. Not all Russian Jews like Israel when they get there, for a variety of reasons, and they then want to go back. Unfortunately, there is a lot of red tape involved in getting back into the U.S.S.R., too. There are inevitably a lot of Jews in Vienna.

A second major international migration within Eurasia is the movement of people from south European and east Mediterranean countries to the Common Market countries. They go to work. Despite all the refugees from the communist lands who have migrated into western Europe, there is still a tremendous shortage of labour inside the Common Market. The labour shortage is particularly marked in certain industries, chiefly construction, transportation, hospital maintenance, coal mining, and general engineering. Part of the reason for the shortage is that these industries have a reputation for not paying as much as other industries, and so Europeans tend to avoid them. The solution has been to recruit cheap labour from outside the northwest European arena. The industrialists of Germany, France, Belgium, and the Netherlands (for they were the main growth countries) opened recruiting offices in countries such as Turkey, Greece, southern Italy, Spain, and Portugal. Generally they hired young single men on 3- or 4-year contracts. There were all sorts of social prob-

"Free the Russian Jews! Free the Russian Jews!" chanted the orderly crowd outside the Soviet Embassy in Ottawa. Daniel Goldberg had come out of a sense of duty; he thought that Jews the world over ought to help one another, and if his presence in the demonstration helped any Jews in Russia who wanted to get out, then he was happy. But his heart was not in the demonstration. He had seen too much violence and experienced too much politicking in Germany in the 1930s to be satisfied that the good of his people lay in that direction. Nevertheless, he continued to chant, "Free the Russian Jews! Free the Russian Jews!"

Solomon Ashkenaze looked at his wife. "We shall not like it here," he said. "The apartments are small and dusty, the labour is hard in the fields, there is danger of Arab attack at all times, and the social services are not so good as in Kiev." His wife nodded in agreement, even though she realized that their dream of many years, a dream of freedom in the new Jewish homeland, was crumbling about them. But they missed their friends back in Kiev; they missed the routine of their old lives. They were not pioneers.

From the back of the line, the office looked remote and distant. Carmelo resigned himself to a long wait. But he had underestimated the efficiency of the Germans; the line moved quicker than he expected. Within several minutes he was going through the door. "Name, last name first," called a young official seated at a desk. Carmelo replied hesitantly, not expecting the young German to understand Italian speech. But the young German gave him clear directions to move along to the next room, where a doctor was to examine him. At the same time the German official stamped a number on Carmelo's inner arm. "Do not remove the number," he instructed. Carmelo shrugged and moved on. The man behind him moved in. "Name, last name first," Carmelo heard the official say, but he was himself then entering the next room. His journey to a German factory had begun.

lems caused by these "guest workers" (which is what they are officially called). Nevertheless, the growth countries continued to recruit guest workers well into the 1970s, because their economies were so dependent upon this supply of labour. Only in 1974 did the growth countries begin to slow down their recruiting campaigns, but then 1974 was a slow-growth year in Europe.

A third major international migration within Eurasia is the movement of Jews from all over the land-mass to Israel. We have examined the movement of Russian Jews to Israel, but we need to note that Russian Jews are not the only Jews. In fact, Jews from all over the world migrate to Israel, some because they were deported from their original country, and some because they wanted very badly to leave. Even before Israel became a state in 1948, there were Jews from all over Europe trying to settle there. (As you know, Israel was then called Palestine.) The Arabs who were in Palestine already, and the British who were supposed to look after the area for the good of the inhabitants, tried to keep the Jews out. Indeed, they tried very hard, even resorting to armed force. But the Jews kept coming. By the end of World War II there were probably more than half a million Jews already in Palestine, and when the U.N. proclaimed the State of Israel in 1948 the new Jewish government set out to attract more Jews to Israel from all over the world. There are now about 3.5 million people in Israel, most of them there voluntarily. They came from Arab countries, such as Jordan, Syria, Saudi Arabia, and the lands of North Africa,

Netherlands Tourist Office

In an Indonesian restaurant in Amsterdam.

which all encouraged the Jews to leave. They came from Europe, where they had been terrorized during World War II .They came from the U.S.S.R., and they came from North America.

A fourth major international migration has been the persistent movement of Chinese into southeast Asia. For centuries the movement was very slow, but it has speeded up in the last 100 years. Chinese people moved in as merchants, a role they still fulfil. They now count for nearly 50% of the population of Malaysia, nearly 80% of the population of Singapore (the key trading town of southeast Asia), nearly 20% of the population of Thailand, and lesser proportions (3 to 4%) of the other countries in southeast Asia (which are named . . . ?).

Other important movements of people within Eurasia, but still across international borders, include the following:

- the Japanese business invasion of Korea and Taiwan. Perhaps as many as 2 000 000 Japanese now live in these non-Japanese countries.

- the Indian transplants in Malaysia. Moved there by the British a hundred years ago to work in the Malay rubber plantations, the Indians are still there (at least their descendants are), and they now form about 10% of the Malaysian population.

- the India-to-Pakistan and Pakistan-to-India movements. When India gained its independence from the British in 1947, and Pakistan was formed at the same time, there were nearly 20 000 000 people displaced by the new states. About 9 000 000 Moslems moved out of India and into Pakistan, and about 10 000 000 Hindus went the opposite way.

- the return home of colonialists. Nearly 500 000 people of Dutch and mixed ancestry returned to the Netherlands when (and after) Indonesia became independent in 1949. The Dutch are still assimilating this large number. (Don't forget, the Netherlands is a small country, with the highest population density in Europe already.) Similar returns were made by French people, including people of mixed ancestry, from what used to be called Indo-China after the French were thrown out by successful terrorist action. There are many former Vietnamese residents in France, for example.

- the movement of Indians and Pakistanis to Britain, looking for work. There are probably about 1 000 000

Cho Kee Chow looked out of the window of his store and saw the rain. It rained a lot in the Malay peninsula, he thought. He turned around to continue serving his customers. He would be sad when dusk fell, for that would mean that another day was nearly over, and he liked serving people in the general store. His family had been serving people for over 100 years in the same store, and he was proud of his job. "Most certainly," he said, "and a half-kilo of rice as well?"

Novosti Press Agency

One of the pioneer old-timers of Siberia.

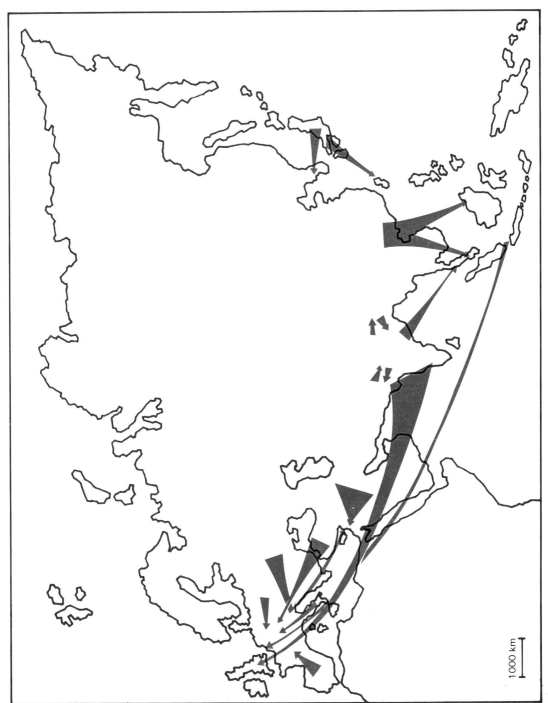

1000 km

Figure 14-1.   Major International Intra-Eurasian Migrations

of them, usually in the cities of Yorkshire, Lancashire, the Midlands, and the southeast (chiefly London).

2 Migrations from Eurasia to elsewhere in the world

The Europeans are undoubtedly the chief contributors to this type of migration. Figure 14-2 shows where most of the Europeans went — and are still going; the movement has by no means stopped. It has been going on for so long and in such numbers that there are probably as many Europeans outside Europe now as there are still in Europe. If you count the people of Latin America, North America, Africa, Asia, and Oceania, you will find about 500 to 600 million people of European and part-European origin, which is approximately the same as the number actually in Europe.

The chief destination of most of these migrants has been the Americas, both North and South. South America attracted mainly the Latin Europeans, although north Europeans (especially Germans) also went in large numbers. North America was the chief destination for north Europeans, but Latin Europeans have also come to the

A hundred years ago the countries of North America used to advertise throughout Europe for immigrants willing to settle in the New World. Cheap transportation and virtually free land were offered. The advertisements were the usual ones of those days: large posters, billboards, handbills, and newspapers.

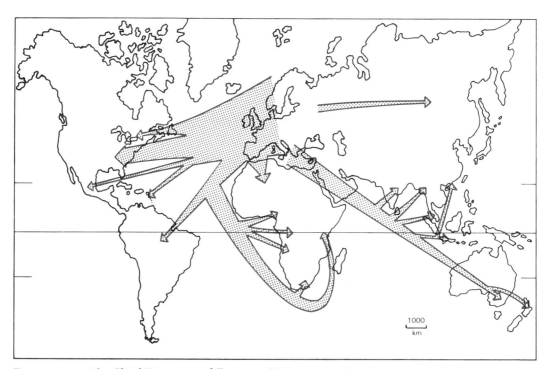

Figure 14-2.   The Chief Directions of European Migrations to Non-European Areas

North in large numbers. All parts of Europe contributed to this massive migration, sometimes one country leading, sometimes another. On the whole the chief source of migrants has been the U.K., making up as much as 25% or so of all the European exodus. Other leading suppliers of migrants have been Italy and Germany, followed by Spain, the Netherlands, Portugal, Greece, and the Scandinavian countries.

Second to the Americas as an attraction to European migrants has been Oceania, especially Australia and New Zealand. Again, all areas of Europe have contributed to this migration. Africa used to rank third as an attraction to European migrants, but various nationalistic policies pursued by several of the new African governments have severely curtailed the flow of Europeans to Africa. Indeed, the movement is now in the reverse direction, as Europeans in South Africa, Rhodesia, Mozambique, Angola, and so on decide that Europe may offer a more secure future.

Why is that?
What is happening in Africa?

Asians have been much less prone to migrate from Eurasia than Europeans have. China and India together have lost perhaps only 30 to 40 million migrants out of their vast domestic populations, compared with Europe's hundreds of millions. The chief attraction for Asian migrants has again been the Americas. Most of the migrants came voluntarily (and still do), but in the early days, about a hundred years ago, many were brought across the Pacific against their will to work on railway construction. Word of opportunity spread back to China, and for a time it looked as though there might be the beginnings of a massive migration across the Pacific. However, both the United States and Canada applied rather harsh restrictions on Chinese migration for many years, eased only since World War II. Japanese migration has been a little different, in that it was carefully sponsored by the Japanese government. Prospective migrants, wives included, were carefully selected by the government and sent off in colonies to the Americas. Some went to Latin America, mainly Brazil and Peru, but most came to North America. Again, the two North American governments set up some barriers to this flow, eased only after World War II.

The flow of Indians out of India was initially to East and South Africa, where there were opportunities created by the British. Indians were in fact transplanted by the British to many areas of their former empire in the tropics, chiefly because the Indians were available in large numbers (which local populations often were not), and because they understood English (which local populations usually did not). The British thus made use of the Indians, and in consequence many Indians found themselves — against their will — in Malay rubber plantations, in Kenyan and Ugandan railway gangs, on Kenyan and Ugandan tea or cotton plantations, in South African shipyards, and so on. When the British withdrew from their empire, the Indians remained. They had worked hard for the British and had usually gained a measure of comfort in their lives. However, not all newly independent African countries welcomed the Indians staying on. Uganda, for example, actually expelled them all at very short notice. Many of these Ugandan Asians, as they were called, came to Canada eventually (in 1972-73).

In recent times other Indians and Pakistanis have come to Canada directly. Some even came in the guise of tourists, determined to stay but not willing to go through immigration procedures. Canada's open-door policy was therefore changed a little in 1973 to make it necessary for prospective immigrants to apply before they left their home countries. The movement continues, but a little more slowly now.

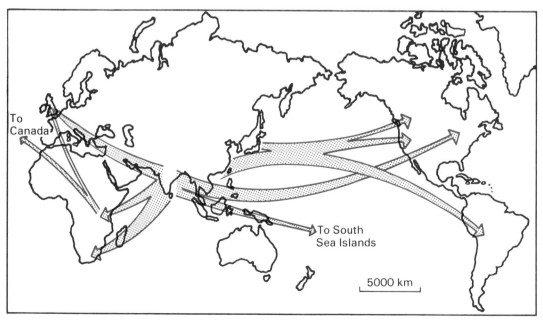

Figure 14-3.   The Chief Directions of Asian Migrations to Non-Asian Areas

3 Migrations from elsewhere in the world into Eurasia

Not many people have migrated to Asia, but a lot have migrated to Europe. They have chiefly come from Africa and the Caribbean, looking for better economic opportunity. Many of the reasons that we examined for the Common Market's desire to recruit relatively cheap labour from Turkey, Spain, etc., also apply in this migration of people from Africa and the Caribbean. The precise link — apart from job opportunity — was the former existence of European empires. For example, the French empire in North Africa was a direct link between Algerians looking for work and the French looking for workers. Thus, since World War II there has been a massive migration of Algerians to France; indeed, it is estimated that one in 7 Algerian men has been to France to work. The work is mostly in construction, and the men save their earnings to take back with them. Naturally, in trying to avoid spending too much of their money inside France itself, many Algerians live very close to a poverty level. Often, they live in shanty towns around the bigger cities. They don't want to rent more expensive homes — otherwise, what was the point of coming to France to work? After all, they are in France for only a short time.

Novosti Press Agency

The centre of a new town in northern Siberia, inside the Arctic Circle.

Much the same background also applies to the movement of Caribbean people to the U.K. The former existence of the British empire meant that English-speaking people were available in the Caribbean. If they would work for low wages in Britain, then so much the better for Britain. Since the alternative to low-wage work in Britain was often no work at all in the Caribbean, the migration to Britain started (after World War II). British firms and government agencies opened recruiting offices in the Caribbean islands, and shiploads of young men and women started to come to Britain. Work was available in the transportation industry and in the hospitals (the men would be station cleaners and bus conductors, the women nurses). There are possibly 1 000 000 Caribbean people now in Britain, and unlike the Algerians, they generally intend staying in their new homeland.

Britain also attracts migrants from other parts of its former empire, though not necessarily for low-wage work. Australians and New Zealanders are fairly numerous, as are Canadians and Americans. Indeed, Britain's most prestigious newspaper, *The Times*, is owned by a Canadian living in London. Other Canadians in England work in

In England, a bus conductor collects the fares. The fares are all different, being based on the distance travelled, and so the conductor has to run up and down the stairs of the big double-decker buses, making change and issuing tickets. The driver just drives.

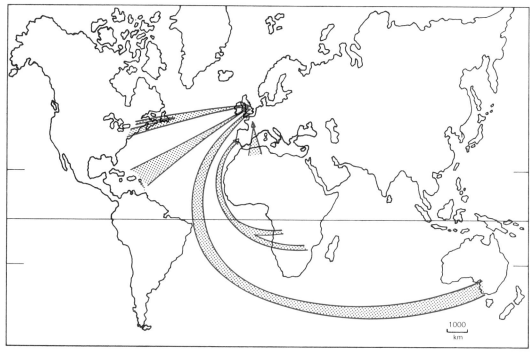

Figure 14-4.   The Chief Lines of Non-Eurasian Migration into Eurasia

TV and radio, own media chains, run shipyards, and act as typists, teachers, and thespians. (What is a thespian?) There are also some rich Americans who think that life is quieter in England.

INTERNAL MIGRATIONS

Large-scale movements of people occur entirely within nations as well as across international frontiers, and it is to these that we now turn. The reasons for movement are often very similar to those for international migration — people seek better economic opportunities elsewhere. However, internal transfers of population may also be generated by the desire of the central government to colonize and populate an area. A government may want to do this for a couple of reasons: the area has potential value to the state if developed; or if the area is left empty, some other country may be tempted to move in. Let's have a look at both of these major reasons.

1   Transfers for economic gain

It is quite normal for some parts of a country to offer better job opportunities than other parts. Not all parts of a country are equally endowed with resources, markets, capital funds, and labour supplies. Those parts that already have these things tend to grow faster than the rest of the country, and so they attract population from the have-not areas. This does not always produce a decline in population in the have-not areas, for the natural increase in those areas may be high enough to support an outward movement of people without causing any decline in numbers. But it may produce a decline. In any event, the areas that already have large populations because they already have many advantages tend to get even more people. Growth is like a rolling snow-ball — it is *cumulative*.

Up to a point. When that point is reached then it *may* work the other way. People will get in each other's way, trains will be overcrowded, housing will be very expensive, schools will be packed, crime and pollution may increase. Growth will grind to a halt.

There are two important examples of this type of population transfer: Italy and Britain. In the case of Italy, economic opportunity lies in the north of the country, especially in Milan and Turin. There are the giant motor factories, the appliance factories, the construction jobs, the transportation and trade jobs, the financial jobs — in other words, the opportunities. In the south — the so-called *Mezzogiorno*, because of the strength of the noon-hour sun — economic opportunities are very limited. Farming is poor, industry is almost non-existent, trade and commerce are slight. So the people move north, in trainloads. The Italian government is trying to slow down the transfer by attempting to create jobs in the south, but it is a slow, slow task.

In Britain, on the other hand, the economic opportunities lie chiefly in the south of the country, and there is a corresponding *southward drift* of the population (see Figure 14-5). The south offers factory work in the expanding sectors of the economy (cars and other engineering), while the north chiefly offers factory work in the declining sectors of the economy (mainly textiles). The south also offers more tertiary (or "service") jobs, because there are more people there already. Thus there are more opportunities in transportation, hairdressing, teaching, nursing, selling, advertising, and so on. In Britain this drift south is gradually returning the population distribution to what it was before the industrial revolution, for during the industrial revolution the opportunity was mostly in the north and people went north looking for work.

2   Transfers for colonization

In these examples, the transfers of population are largely into relatively empty lands — lands that offer opportun-

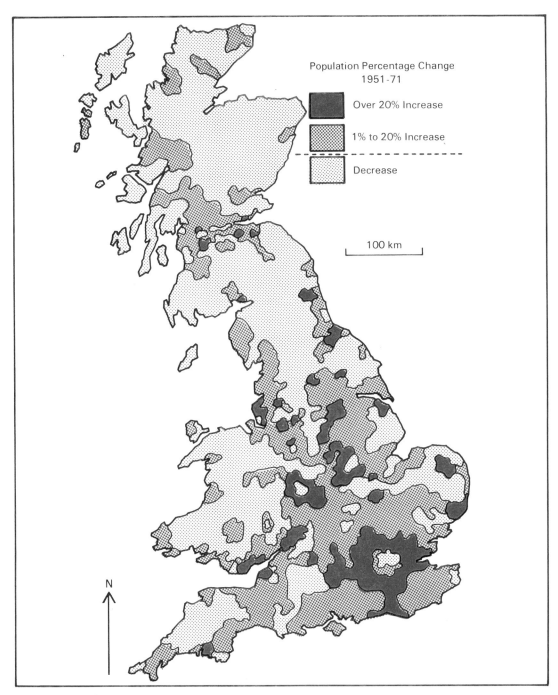

Figure 14-5.   The Population "Drift South" in Britain

ity for development, or lands which the state wants to claim as its own. Probably the most famous example is Siberia. For centuries, various Russian governments used it as a place to exile unwanted citizens from western Russia. During these centuries these exiled prisoners were responsible for a little development of the area — some road and rail construction, town building, mining, and subduing the native peoples who resented their presence. During the period since the communist take-over of the U.S.S.R., this random sort of development has been gradually replaced by a much more determined effort to establish firmly Soviet rights in Siberia. Large incentives (money, cheap housing, promises of promotion) are now given to people willing to go and work for extended periods in Siberia, and the use of Siberia as a punishment area has now almost entirely gone. The Russians are determined to colonize Siberia, so that it becomes not only a source of economic wealth to them, but also a protective bulwark against any possible invasions from China. The Russians remember Genghis Khan, who was not Chinese but who symbolizes to the Russians the threat from the east. The Russians also remember Napoleon and Hitler, and so they additionally value Siberia as a possible escape-in-depth when faced by any attack from the west. It is estimated that about 25 000 000 of the 30 000 000 people in Siberia have their origins in the western U.S.S.R. Most of them have moved into Siberia fairly recently, and they represent yet another major migration of mankind.

Over a shorter distance, but still important in the numbers of people involved, is the Chinese migration into Manchuria. The Chinese have trickled into Manchuria for centuries, but even as recently as the 1920s they still formed a minority in comparison with the local Manchus. Since then, and especially since World War II, there has been a massive government-sponsored drive to colonize Manchuria, which now houses some 50 000 000 to 60 000 000 Chinese compared with only about 2 000 000 to 3 000 000 Manchus. It is partly this Chinese drive into Manchuria, which is right next to the eastern parts of Siberia, that has made the Russians so anxious to colonize Siberia. Also important in making the Russians jittery about Siberia has been the Chinese drive to populate and develop their western provinces of Tibet and Sinkiang as well. You can see their locations in Figure 14-6.

On a still smaller scale is the Japanese push into Hokkaido. Hokkaido is the most northerly of the 4 major

The very mention of Siberia was often enough to send a shiver through the hearts of potential rebels!

Siberia now welcomes people warmly!

Novosti Press Agency

Part of downtown Khabarovsk in eastern Siberia.

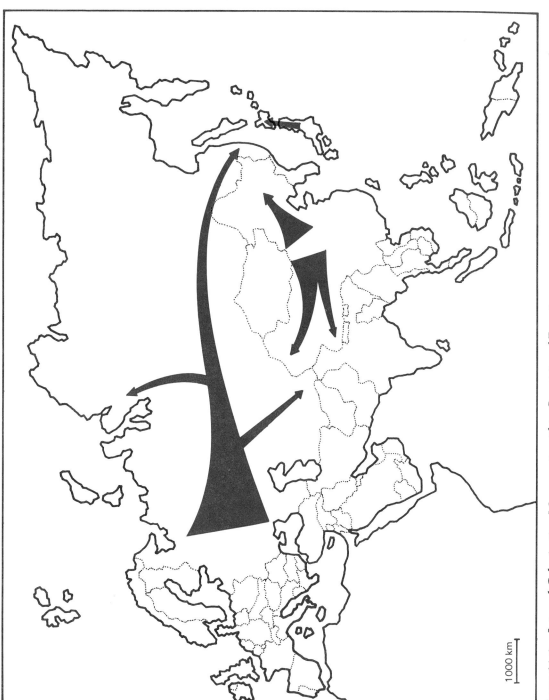

Figure 14-6. Internal Colonization Movements within Countries of Eurasia

1000 km

Japanese islands. Unlike the others, it is largely unsuitable for rice cultivation. In fact, Hokkaido has been so different from traditional Japan that the Japanese for centuries left it as the retreat of the Ainu people. The Meiji Restoration was responsible for changing that attitude. After 1870 the Japanese government began a sustained move to colonize Hokkaido. Railways were built, and farmers were moved in. They had to develop entirely new ways of farming, because rice would not grow there; but dairying and fruit growing soon became important. From a population of just a few thousand (mostly Ainus) in 1870, Hokkaido has grown to a population of about 10 000 000, nearly all Japanese.

The Ainu are something of a mystery. They are white-skinned and have more body hair than the native Japanese. They have occupied Hokkaido throughout history, and may at one time have spread even farther afield. Their numbers have never been great, however — perhaps never more than 100 000.

SEASONAL MIGRATIONS

Some migrations are only temporary; people move to an area, and after a season they move back to their original area. These seasonal migrations are thus different in character from the migrations we have so far examined, wherein the purpose was permanent settlement in a new area. In the case of seasonal migrations, the purpose is to return to home-plate, as it were. There are several different reasons for seasonal migration, including economic, survival, and leisure reasons. Let's take them in turn.

1   Seasonal migrations for economic reasons

The 2 main employers of seasonal labour are agriculture and the tourist trade. Agriculture uses a lot of labour during the harvest, especially if the crops are tender and need a great deal of hand care (such as soft fruit). Thus at harvest time, workers are often brought in from wherever there is a surplus of people looking for work. It may be from the local towns, or it may even be from another country. Examples include the migration of southern French people to the sugar-beet fields around Paris; the migration of Algerians into the French vineyards; and the migration of southern Italians into the Lombardy rice fields. In most cases the migrants return home after the harvest. Indeed, in Luxembourg migrant workers are requested to be out of the country before December 15.

The tourist trade is usually a summer attraction, but there is an increasing demand for labour in certain areas even in winter, as the winter sports industry grows. During the summer (or the winter, in certain areas) labour

The movement of seasonally migrant labour of this type is gradually diminishing as the farmers take more and more to the use of machinery, even for delicate work. For example, grapes can now be harvested by machines, although not many of the best ones yet are.

moves into the tourist areas from towns all across Europe. Most tourist-trade labour is urban-based. Some of the workers are students, some are regular hotel workers seeking a change, and some are just people looking for work. Perhaps the largest of these movements is the seasonal migration of Italians to Switzerland, but the movements into the Mediterranean coasts of France and Spain are also very large.

2   Seasonal migrations for survival reasons

It has been the practice of certain groups of people to wander from place to place in search of the means to survive. Usually for these people, their environments have been harsh — either too dry, as in the Middle Eastern deserts and Central Asia, or too cold, as in Lapland. There are several different tribal groups involved in these movements, and the patterns of movement are not precise and regular. The migrations take place strictly on the basis of need; the very survival of the people is at stake. Thus the timing of the migrations is erratic. (Why move on if you don't need to? Why stay if you need to move on?) The only fixed considerations are the destinations, be they water holes in the deserts or reindeer-grazing areas in the tundra. Without these means to survival the peoples would perish. In Scandinavia the Lapps migrate through Lapland, which runs across the north of the Scandinavian peninsula, from Norway in the west to Russia in the east. In Siberia the Tunguz have a similar tundra-located migratory life-style. In Central Asia the Kazakhs and the Kirghiz migrate seasonally in search of goat, sheep, and camel pasture. In the Middle East the Bedouin migrate through the deserts of Jordan, Syria, and Saudi Arabia.

In all cases the migrations have been the subject of government pressure. Governments do not like people wandering about so much. People who roam about the country are difficult to govern. Not only that, but when the migrations cross international frontiers (as many do) they are even potentially dangerous; there may be border incidents to say the least, and there is every likelihood of smuggling. The governments have almost won, but not quite yet. *Nomadic herding* still survives as a basic life-style.

One very special type of seasonal migration for reasons of survival is *transhumance*. This used to be extremely

Per Bjorling looked at the line of posts. Soon there would be a wire fence, he thought. Why, it's not so long since there was just the occasional marker to show the location of the border, and now look at it! He could even remember that his father had told him when he was a child that in those days there were no markers of any sort. The frontier might just as well have not existed as far as his people were concerned. They could follow the reindeer herds at will in those days, but now it was difficult even for the reindeer to travel across their old grazing grounds. Things were really getting too organized!

common at one time, but it too has largely (but not entirely) died down. The idea behind transhumance is that mountainous areas offer valuable pasture for cattle, sheep, and goats during the summer, even though they cannot be used for farming at all during the winter. Rather than "waste" the pasture that is freely available in summer, people drive their animals up the mountains in spring to use the pasture. Meanwhile, down in the valleys and on the lowlands, other people remain behind to grow as many crops as they can while the animals are out on the hills. Some of these crops are stored after harvesting for feeding to the animals when they are brought down from the hills in fall. In this way, free pasture is provided for the animals during the summer, and at the same time extra land is available in the valleys for growing crops, because the cattle have been moved out. During winter more animals can be supported than would otherwise be the case because of the extra feed that was grown during summer. The practice has obvious advantages, and it has clearly not died out completely. What has happened is that the animals often still move uphill in spring and down again in fall, but only under the care of a few specialized farm-men. It's the seasonal movement of large groups of people that has almost died out, not the principle of transhumance itself. You can therefore still see it practised in most mountainous areas of Eurasia, from Norway to Spain, and from the Pyrenees to the Himalayas.

3   Seasonal migrations for leisure

The tourist trade is now one of the world's largest industries. Millions of people travel for a few weeks to different places in the world, and then return home. In Eurasia the top tourist countries are shown in Figure 14-7; you can see that the chief recipients of tourists are Spain, France, Italy, and the Scandinavian countries, followed by Austria, West Germany, the U.K., and Switzerland. It is also quite apparent that within Eurasia the chief country from which tourists come is West Germany. Outside Eurasia, it is the U.S.A. Figure 14-8 is an attempt to show the relative importance of the volume of tourist traffic pouring into the different countries. It is obvious that Europe feeds on the tourist trade, as well as feeding it. Asia, by contrast, has hardly any tourist trade, although Hong Kong and Japan do have a little.

Millions of Tourists *to*:		Millions of Tourists *from*: (leading country only)	
Austria	9.0	West Germany	5.5
Belgium	4.0	U.K.	1.0
Bulgaria	2.5	Turkey	0.6
Czechoslovakia	3.5	East Germany	0.3
France	14.0	West Germany	2.0
East Germany	5.0	West Germany	2.0
West Germany	8.0	U.S.A.	1.5
Greece	1.5	U.S.A.	0.3
Hong Kong	1.0	U.S.A.	0.3
Hungary	6.5	Czechoslovakia	2.6
Italy	13.0	West Germany	3.0
Japan	1.0	U.S.A.	0.3
Lebanon	1.0	Syria	0.4
Netherlands	2.5	West Germany	0.6
Poland	2.0	East Germany	0.6
Portugal	3.5	Spain	1.7
Romania	2.5	Czechoslovakia	0.7
Scandinavia	11.5	West Germany	9.5
Spain	22.5	France	9.0
Switzerland	7.0	West Germany	1.6
U.S.S.R.	2.0	Poland	0.4
U.K.	7.0	U.S.A.	1.6
Yugoslavia	5.0	West Germany	1.2
for comparison:			
Canada	38.0	U.S.A.	37.0

Figure 14-7. Tourist Data for All Eurasian Countries Which Receive at Least 1 Million Tourists per Year

The reasons for the predominance of Europe in the Eurasian tourist business are mainly concerned with the greater degree of economic growth achieved by the European countries (compared with the Asian countries). Higher wage rates, permitted by higher productivity, enable the workers of Europe to take time off from their jobs for at least 2 or 3 weeks every year. Many use this opportunity to seek the sun, because for the rest of the year they usually work indoors. Accordingly, the reliably sunny areas of the Mediterranean coastlands are very popular tourist destinations. They usually go by the name *riviera* (French Riviera, Italian Riviera), but in Spain the name *costa* is common (Costa Blanca, Costa del Sol). Along all these coastal areas throughout the summer there is a steady stream of winter-pale people one way and sun-tanned people the other.

A growing part of the tourist trade is the winter sports business. Since World War II especially there has been a spectacular increase in the leisure use of snow and ice. Throughout the Alps (chiefly) there are now numerous

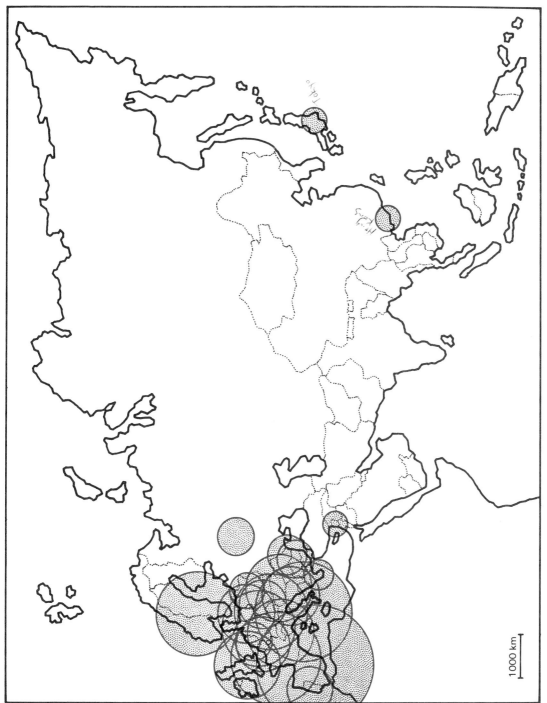

Figure 14-8.   All Countries in Eurasia Receiving at Least 1 Million Tourists Visitors per Year

resorts that come alive only during the winter, catering to skiers, skaters, bobsledders, curlers, and so on. Wengen in Switzerland and Kitzbühl in Austria are just two examples out of many.

Would it be true to say that winter sports have snowballed?

CONCLUSION

Do you get the impression from reading this chapter that there are millions of people wandering about the face of the Eurasian continent? There are, indeed! That is precisely what migrations involve. As you now know, some of the migrants move once and then settle, but some of them are constantly on the move. Others move repeatedly but not constantly. Can you give examples of these 3 different sorts?

Another point about migrations — apart from their size and scope — is that they are the major means by which population is distributed over the earth. Even now, the population distribution you can see on a map (see Figure 2-1, for instance) is not necessarily the ultimate pattern. Migrations may change it in the future just as they have done in the past.

QUESTIONS AND EXERCISES

1. Write an essay pointing out all the different causes of migration that you can find.
2. What are the different barriers that governments can impose on migrations that cross international frontiers?
3. Research the contribution of immigration to the growth of population in Canada.
4. What advantages do you think the Europeans brought to the areas of the world that they migrated to?
5. What are the disadvantages that people who migrate might find (or feel) in their new homelands?
6. Write an *imaginative* and *reasoned* essay on what you think the world would be like today if there had been no migrations outside Eurasia at all.
7. Who are the gypsies?

15. THE INDIAN SUBCONTINENT

REGIONALISM

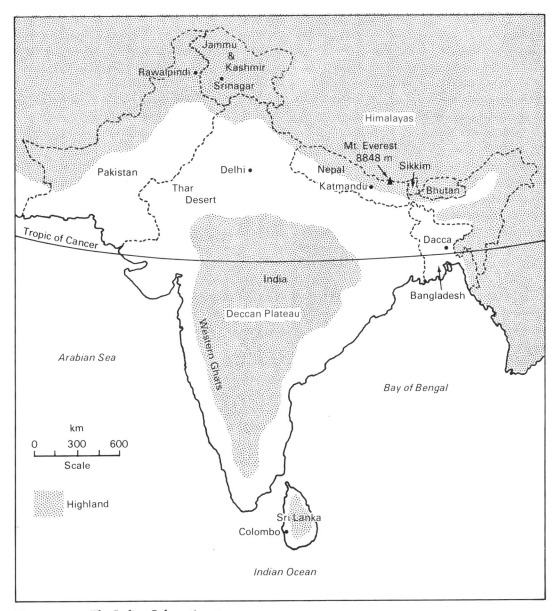

Figure 15-1. The Indian Subcontinent

In case you missed it, go back and inspect Figure 15-1. It shows 8 countries that comprise an enormous land mass. Known as the Indian sub-continent, this vast land mass is very nearly half the size of Canada. That is one fact that makes the region stand out. Another is the enormous number of people who live there — over 700 000 000 of them, and the total increases yearly by approximately 15 000 000. This is an impossible figure to imagine, but we can try. Lying head to toe in a continuous line, that yearly increase in population would stretch for over 26 000 km, or more than halfway around the earth. If the entire population were to march past in single file at the steady speed of 5 km/h, the line of people would never come to an end. The *net* increase in population is about 55 000 people daily, or 2300/h, or 38/min. In the time it took you to read this, can you work out how many people were added to the population of the sub-continent? About 30!

The modern history of the region dates from the end of the Second World War. For years the sub-continent had been a possession of Britain. In August 1947, following many years of unrest in India, and especially civil disobedience as advocated by Mahatma Gandhi, Britain resolved to partition the region on religious lines. The rule was transferred to independent national governments in Pakistan (Moslem) and India (predominantly Hindu). Following partition, much violence took place, including massacres, on religious grounds, and relations between India and Pakistan were never very good. A complicating factor was the country of Jammu and Kashmir. Although

Hinduism is the most important religion in India. It is a loosely organized religion, involving the worship of many gods. Worship takes many forms, but there are some common threads. The chief god is *Vishnu*, representing the hope of salvation. Another important god is *Shiva*, representing destruction. Cows (not buffaloes or oxen) are regarded as holy, and a devout Hindu will try to die holding the tail of a cow, as this is the only animal reputed to be able to swim across the sea of blood that appears to everyone when they die. Hindus hold the River Ganges as sacred, and try to make a pilgrimage to it as often as possible. On its banks stand many temples. Dead bodies are sometimes cremated there, too. Reincarnation is another Hindu belief.

Although Kashmir is a trouble spot, a town called Gulmarg in the country is a winter sports resort. On the ski slopes, where snow lasts from December to March, only 20 people a day used the facilities in the first season. Not bad if you like a mountain to yourself!

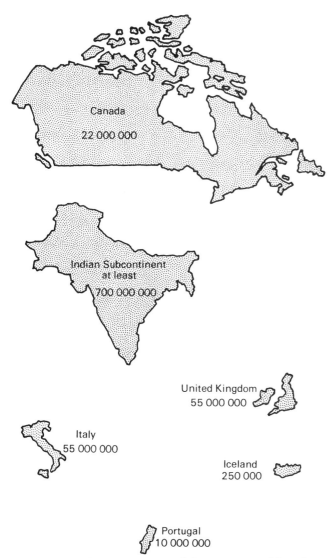

Indian Government Tourist Office

The Ganges is a holy river, sacred to Hindus. Here pilgrims are seen bathing. Can you see the holy cow?

Figure 15-2.   Some International Comparisons of Population

Indian Government Tourist Office

Market day in the village is a bustling, open-air affair. People haggle over prices, meet friends, and generally treat it as a social event. Later in the day, they will return to the smaller, outlying villages from which they have most likely walked.

its population is mainly Moslem and it was thus allocated to Pakistan at partition in 1947, the state blocked the entrance of Pakistani troops in that year and the ruler of the country advocated union with India instead. (The ruler was a Hindu.) Pakistan objected, not only because the Moslem majority in Kashmir had been denied a chance to decide whether they preferred union with Pakistan or India, but because the country contains the headwaters of the Indus River, vital to the Pakistan economy. India was now claiming sole right to those headwaters, while Pakistan claimed at least an equal share in

Country	Total land area km²	Estimated 1974 populations	Per capita $U.S. national income (1970)
Bangladesh	141 123	85 000 000	n.a.
Bhutan	47 000	900 000	70
India	3 233 280	580 000 000	89
Kashmir	207 360	5 000 000	n.a.
Nepal	140 800	11 000 000	80
Pakistan	803 880	70 000 000	132
Sikkim	7 110	200 000	n.a.
Sri Lanka	64 740	14 000 000	160
Canada	9 976 140	22 000 000	3 769

n.a. — data not available

Figure 15-3. The Populations of the Countries of the Indian Subcontinent

Indian Government Tourist Office

This peaceful mountain scene of Kashmir changes abruptly with the seasons. At what season do you think this photograph was taken?

them under international law. At the moment, the status of Jammu and Kashmir is still in question.

Up until December 1971, Pakistan had been a split country — part on the east of India, part on the west. Then, after civil war and armed intervention by India, East Pakistan became independent and was renamed Bangladesh. Although still mainly Moslem, as is Pakistan (formerly West Pakistan), reunion is most unlikely.

One of the more significant facts to remember when dealing with this region is that the vast majority of the inhabitants live on the land, farming. While their living standards are not very high, it is very difficult to go to the cities to find work. Most of them stay on the land and try to farm as best they can. In this venture, the dominating influence is the weather. Most of the Indian subcontinent is affected to some extent by the *monsoon*.

The monsoon is a seasonal wind that blows *off the land in winter*, so that it is dry. But it blows *off the sea in summer*, so that is responsible for heavy rain at that season. Turn back to the chapter on climate to discover some more facts about the monsoon.

The crop that dominates the economy is rice. Rice is a *staple cereal* in this region, and try as they will, people can never grow enough of it. The main variety of rice is paddy rice. This method of cultivation requires that the land be flooded to a depth of 14 cm. Low dikes enclose each field to prevent the water flowing away, and in the areas with heavy monsoonal rainfall, daily storms keep the levels of water in the fields topped-up. This is vital, for in tropical countries water evaporates quickly.

There is some evidence that the harijans are forming political action groups in India; in 1974 there were riots between harijans and higher castes in Bombay.

When the monsoon begins in India sometime in June, the farmer sows rice seeds in a nursery field to germinate and become seedlings about 25 cm tall. Then these are transplanted and carefully set in rows so that each plant is separated from the next by perhaps 20 cm. Transplanting is usually complete by mid-July to mid-August. Then the farmer has to weed and apply fertilizer from time to time. By December, rice planted in early June is ready for harvest. This lasts until mid-January in some cases. The cut grain is allowed to dry for a day or two before it is brought back to the village. Then it is threshed by hand or by oxen treading the grain out of the husks.

In a few favoured areas, a second crop of vegetables such as peas or beans can be grown if the ground is still damp enough. These are all harvested by March and then the land and its people wait for the June monsoon again.

Wherever the land is not favoured with heavy rain from the summer monsoon, such as in Pakistan, irrigation is practised. Rather than grow rice, which requires large amounts of water, the Pakistanis have concentrated on wheat. To this end the waters of the Indus have been impounded by a series of dams and barrages so that millions of hectares can be irrigated. One of the more important dams is the Lloyd Barrage at Sukkur, which sends water to fields up to 320 km away, and covers an area in Sind Province of perhaps 20 000 km^2.

Another area that requires extensive irrigation is the Deccan plateau of southern India. Here the full force of the summer monsoon is rarely felt. It is necessary to store water in wide, shallow ponds called *tanks* in order that life may continue in the dry winter and spring. The staple cereal here is millet. Since tanks are utilized for every purpose, including drinking water, and losses of water are heavy during the hot dry winter and spring, the Indian government has been trying to have farmers rely more on *tube wells*. Perforated pipes several centimetres in diameter are forced and drilled into the earth to tap the water table. Electric pumps deliver a reliable supply of water to the fields. Of course, such a system is only available to richer farmers.

Really, India does not suffer from a shortage of water at all. It is just necessary to redistribute it somehow. Even though there are now well over 1 000 000 tube wells in operation, there is still too much reliance on the monsoon. So a $4.5 billion program to remove excess water from the Ganges during floods would channel it more than

Indian Government Tourist Office

A Kashmiri shepherd drives his flock along a country road.

Mick Horner

A Pakistani ox-cart.

Indian Government Tourist Office

Tube wells like this are very important to India.

3000 km south to the villages of the Deccan. This is expected to increase irrigated land in south and central India from 7 800 000 ha to 18 500 000 ha. It is expected that the project will be completed by the end of this century.

There is urgent need for the countries of the region to learn to protect themselves against the uncertainties of weather, as the newspapers of the late 1960s showed. India, for example, had failures of the monsoon in 1965, 1966, and 1972. In 1973 millions of Indians died as a result of severe malnutrition. Even in a good harvest year, there is barely enough food to go around.

Pakistan is hardly in any better position. In 1972 it too, had to buy large amounts of wheat abroad, chiefly in the U.S.A. A million tonnes of wheat changed hands for $60 million, although the money did not have to be paid in less than 40 years. Canada also contributed about $10 million worth, and on the same terms.

Bangladesh was roused from its euphoria after the December 1971 peace. Normally running a food deficit of 2 000 000 t of grain in any year, the disruption due to the war had resulted in at least 10 000 000 refugees fleeing to India so that they were not available to work their land. The country was short 3 000 000 t of food in 1972. At first it was thought that farmers could substitute rice for other crops, jute perhaps, but this plan was not workable.

On the other hand, sometimes too much rain comes. In Pakistan during August 1973, floods endangered 15 000 villages and affected 11 000 000 people. Tube wells and irrigation canals were badly damaged. Total losses ran to over $500 million, including the destruction of 1 500 000 t of food grains.

It is not merely the weather that has a depressing effect on agriculture. Many parts of the sub-continent are too poor, too badly organized, too ignorant of modern farming practices, too illiterate, and too overpopulated to make much of the land that they hold, even in ideal conditions. Centuries of farming have cut into soil fertility, while over-irrigation has produced many saline and alkaline soils. Perhaps 25% of the land is suffering from soil erosion. Since the vast majority of the population lives on the land, the dominant crops are food grains such as rice, wheat, millet, corn, barley, and lentils. Such crops occupy more than 75% of the cultivated land. Cash crops such as cotton, jute, tea, and oil seeds occupy very little land area.

You must not think that the sub-continent suffers from a lack of farmland, though. On the contrary, as Figure

India has great numbers of university graduates whom it could well use in the villages, where 8 out of every 10 Indians live. However, such villages offer few if any attractions to newly graduated doctors, for example. Out of 125 graduating medical students at a college in Gujarat, 85 applied for U.S. immigrant visas several days later in Bombay. They had chartered a bus to get there. Not only do many Indians emigrate to the U.S.A., but numbers have come to Canada and especially to Great Britain.

Nearly 250 000 Pakistanis are employed by the oil-producing Arab states along the Persian Gulf.

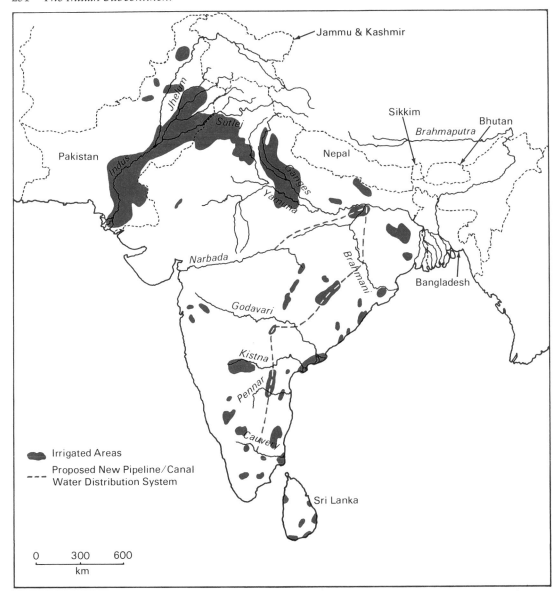

Figure 15-4.   Irrigated Areas in the Indian Subcontinent

15-5 shows, compared with other parts of the world, India and its neighbours are quite well provided for in terms of land.

One clue to the region's markedly poor performance in farming is provided in Figure 15-5. Notice the number of farms that are listed opposite each country. In many places, a farmer splits his land evenly between his sons, so that when he dies, each son will have a means of sup-

Country	Arable land area km²	Population est. 1974	Land under irrigation km²	Number of farms
Bangladesh	90 690	85 000 000	6 860	n.a.
Bhutan	50	900 000	none	n.a.
India	1 646 100	580 000 000	275 200	48 882 000
Nepal	19 800	11 000 000	1 810	n.a.
Pakistan	192 350	70 000 000	125 050	12 154 928
Sikkim	100	200 000	none	n.a.
Sri Lanka	19 790	14 000 000	4 650	1 169 801
Canada	437 670	22 000 000	3 460	366 128
Japan	54 460	108 000 000	2 626	6 056 534
Netherlands	8 480	14 000 000	51	184 613
United Kingdom	72 270	57 000 000	88	395 670
West Germany	80 830	60 000 000	270	1 761 114
Israel	4 170	3 000 000	173	n.a.

Figure 15-5.   Some Land Use Data for the Indian Subcontinent Compared with Other Parts of the World

Indian Government Tourist Office

This girl is at work near Darjeeling. She is picking only the tender young leaves. It will take a long time for her to fill her basket.

port. Over the centuries, this has resulted in the drastic reduction of farm size and the *fragmentation* of holdings. Such "farms" are often incapable of feeding all who depend upon them. Yet large families are still regarded by many people as indispensable. Children provide a workforce, and a source of support in one's old age.

The Ganges plain is very fertile, since it is floored with alluvium brought down by rivers from the Deccan to the south, and the Himalayas to the north. It is nearly impossible to find a pebble in the rich, fertile soils of the region. Villages are small and vary in size between 500 and 10 000 people. Houses have thick mud walls, which are easily eroded in the wet season but just as easily repaired. Most roads are of the mm/dd type: muddy during the monsoon, and dusty during the dry season. Few villages have electric light although there may be supplies for tube wells and irrigation pumps. Cooking fuel is provided by cow dung, which is dried, then burned to supply heat. It is estimated that half India's energy supplies are produced in this fashion! Personal possessions are few in number — perhaps some pieces of wooden furniture; some iron, brass, or copper pots and cooking utensils; and a few articles of clothing.

A great many attempts have been made to pull the economies of these countries out of the doldrums that they seem to have been in for years. India provides an excellent example of the basic problem: How can it be possible for the Indian economy to grow?

To begin with, all economies are divided into 3 sectors. The first, or primary sector, is the production of raw materials — mining, agriculture, and fishing. Sector 2 takes these basic products and converts them into goods such as cars, machines, fertilizers, chesterfields, and so on. Sector 3 industries deal with the production of services, including wholesale and retail selling, insurance, defence, education, car repair, etc. Which sector would best repay investment in it? Jawarhalal Nehru was prime minister of India from 1947 until his death in 1966. He devised 3 5-year plans to boost the production of goods (sector 2 industries). He met with great success in stimulating in-

Indian Government Tourist Office

The cultures of India are many and varied. Here men of the Bhangras, a Sikh sect, perform a traditional dance.

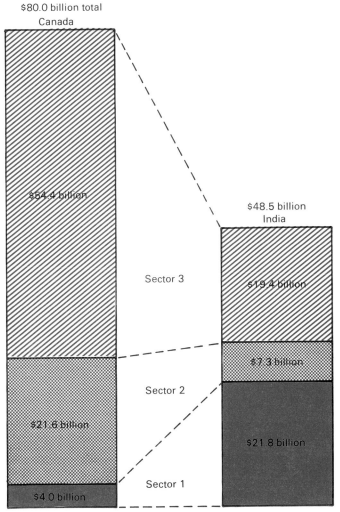

$80.0 billion total
Canada

$54.4 billion

$48.5 billion
India

Sector 3

$19.4 billion

$7.3 billion

$21.6 billion

Sector 2

$21.8 billion

Sector 1

$4.0 billion

Figure 15-6.   **A Comparison Between the Economies of India and Canada for 1971. Note the overwhelming dependence in India upon primary (mainly agricultural) activity. In 1971 the per capita National Income in Canada was $3769; in India it was $90.**

dustrial production, so that India is now capable of making nuclear bombs and jet aircraft, and can undertake almost any kind of research. However, there was little or no benefit for the 70% of the people who live on the land. They did not live in cities and thus didn't work in industry. They had to produce for themselves what they needed to survive by means of subsistence agriculture, which all too often failed them. They could not avail themselves of the new fertilizers or farm machinery; they were too poor. They formed almost a nation apart.

It was not until April 1969, after 2 successive famines (1965 and 1966) that a fourth 5-year plan was put forward. It was intended to raise the standard of living of everyone in the country by supporting agriculture. If the farmers could produce more, they would have a surplus of food to sell. Therefore, they would be able to buy more goods and services and give the rest of the economy a boost. Figure 15-7 shows how the Indian economy depends on agriculture. If the monsoon fails, so does the economy.

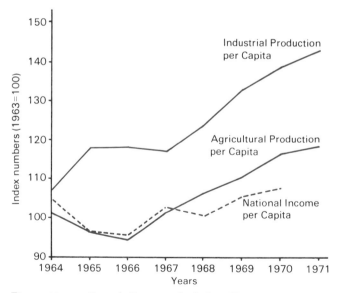

Figure 15-7.   Growth Rates in the Indian Economy.
Note the depressing effect of agriculture upon the whole economy and how industrial growth affected it very little up to the late 1960s.

Starting in about 1968, the *Green Revolution* came to India. Dr. Norman Borlaug, who was later to win a Nobel

Prize for his work, started experimental farms in India. He used new strains of rice and wheat that he and fellow agronomists, botanists, and other scientists had developed in Mexico in what had proved a fairly successful attempt to raise the levels of nutrition in that country. In India, interested and envious farmers clamoured for the new seed and fertilizer but found that the government had restricted subsidies and grants to pay for such things to the larger land-owners. There was not enough for everyone who wanted it. A choice had to be made. But the choice was soon forced by a new trend; the smallest land-owners (over 60% of all the farms in India were less than 2 ha in size) were selling out in droves and moving to the cities. The government quickly extended aid to the small farmers so that they too could afford the doubled or even quadrupled yields of the new hybrid wheat, rice, maize, and millet. For the government knew that there are no employment possibilities for uneducated, unhealthy peasants in the city industries. Unemployment was already high enough in industry, and when cities such as Calcutta and Bombay were inundated by floods of people from the

Indian Government Tourist Office

With ever-increasing numbers of people coming to the cities, most urban centres of India are very crowded. This Calcutta street scene exhibits curious mixtures of the old and the new, as well as the bustling nature of Indian urban life.

WHEAT

		1961	1962	1963	1964	1965	1966	1967	1968	1969	1970	1971	1972
Bangladesh	a	57	62	74	57	53	55	73	78	117	120	126	110
	b	33	40	45	35	35	36	59	59	93	105	112	70
	c	574	636	607	601	647	646	809	758	796	874	888	636
India	a	12 927	13 570	13 590	13 499	13 422	12 656	12 838	14 998	15 958	16 626	18 241	19 163
	b	10 997	12 072	10 776	9 853	12 257	10 424	11 393	16 540	18 652	20 093	23 833	26 477
	c	851	890	793	730	913	824	887	1 103	1 169	1 209	1 307	1 382
Pakistan	a	4 639	4 923	5 022	5 019	5 318	5 155	5 344	5 983	6 160	6 229	5 978	5 797
	b	3 814	4 026	4 170	4 162	4 590	3 916	4 334	6 418	6 617	7 294	6 476	6 891
	c	822	818	830	829	863	760	811	1 073	1 074	1 171	1 083	1 189

a = area harvested ('000's ha)   b = production ('000's t)   c = yield (kg/ha)

RICE

		1961	1962	1963	1964	1965	1966	1967	1968	1969	1970	1971	1972
Bangladesh	a	8 484	8 694	9 008	9 229	9 360	9 071	9 889	9 742	10 314	9 912	9 776	9 500
	b	14 426	13 305	15 935	15 754	15 751	14 362	17 564	17 000	18 008	16 714	14 883	14 250
	c	1 700	1 530	1 769	1 707	1 683	1 583	1 776	1 745	1 746	1 686	1 522	1 500
India	a	34 694	35 695	35 089	36 462	35 273	35 251	36 437	36 966	37 680	37 592	37 334	36 500
	b	53 494	49 826	55 497	58 962	45 983	45 657	56 418	59 642	60 645	63 338	64 102	59 000
	c	1 542	1 396	1 550	1 617	1 304	1 295	1 548	1 613	1 609	1 685	1 717	1 616
Pakistan	a	1 214	1 186	1 286	1 356	1 393	1 410	1 420	1 555	1 569	1 503	1 456	1 483
	b	1 691	1 643	1 788	2 025	1 976	2 046	2 249	3 048	2 384	3 300	3 390	3 356
	c	1 392	1 386	1 390	1 494	1 418	1 452	1 584	1 960	1 519	2 196	2 328	2 263

Figure 15-8. **The Course of the Green Revolution**

land, there was nowhere for them to live except the streets. Hundreds of thousands of people live on Calcutta's sidewalks. And the streets are where many of them die.

Then, disaster struck. In 1972 the monsoon was later and lighter than usual, so that the rice crop failed for many farmers. Although the monsoon was right on time and gave plenty of rain in 1973, that was the year of price hikes for crude oil from the Arab countries. The price actually rose by more than 300%. Crude oil is the base of most modern fertilizers, and the Indian sub-continent has few oil resources. Buying crude oil in 1973 completely wiped out India's foreign exchange reserves. Apart from anything else, this meant that should the rains fail in 1974, then India would be unable to purchase food abroad but would have to rely even more on foreign aid. It was a bleak prospect, and in June 1973, Mrs. Indira Gandhi, India's prime minister, came to Canada to seek out possible lines of credit for emergency food supplies.

We have already noted that India and its neighbours are not greatly industrialized. Even if the factories could be built, the machinery produced or imported, and supplies of raw materials assured, the labour supply would be a problem, even though the unemployment rate is at about 20%. It is *skilled* labour that is required, and this is in very short supply. In addition, the work-force left in the fields must produce just as much food as ever and ship it to the cities. Just how likely all this is to come about in the immediate future can be seen by examining Figure 15-9.

Any examination of the Indian sub-continent should include a mention of industry. However, most of India's industry is concerned with producing the necessities of life. There are not very many luxuries, and a great deal of the utensils, clothing, and other products used on a day-to-day basis are actually produced in the villages and small towns of the countryside. Every hamlet has its weaver and potter, for example. One industry that is likely to increase fairly rapidly in the sub-continent, though, is tourism.

In 1971, more than 7000 Canadians visited India. It could have cost them as little as $440 return jet air fare. The hotels were good bargains for Canadians, too; a good Western-style hotel room cost about $8 single, and Indian-style hotels (good ones) cost $3 to $4 per night, also for a single room. Jets reach India via Europe, and the

The 1973 tripling of the price of crude oil from the Persian Gulf hit India and Pakistan hard, because they lack the foreign exchange to pay for the oil. Besides, they need it to make fertilizer, absolutely essential for use with the new hybrid cereals. In 1974, India alone was short of 600 000 t of fertilizer.

The Taj Mahal at Agra was built as a tomb for his favourite wife by Shah Jehan, who also lies buried there. The building represents a mixture of Indian and Islamic cultures and is a favourite tourist attraction.

Indian Government Tourist Office

	India	Sri Lanka	Pakistan	Canada
Telephones	1 293 000	64 000	220 000	10 253 000
Doctors	112 000	3 242	13 011	31 116
Hospital beds	325 500	36 845	42 609	211 637
Cars	619 000	88 300	150 500	6 976 720
Televisions	49 000	—	99 000	7 610 000
Energy consumption in the equivalent of tonnes of coal	102 520 000	2 070 000	12 880 000	201 380 000
Literacy rate %	27.8	75.3	18.8	98
Birth rate/1000	42.8	29.9	50.9	17.2
Death rate/1000	16.7	7.6	18.4	7.3
Expectation of life at birth (years)	41	61	50	71

Note: To compare countries accurately, the *populations* of the countries must also be considered. (Figure 15-5)

Figure 15-9.   Some Indicators of the Quality of Life, 1971

Indian Government Tourist Office

These houseboats on a Kashmir lake are for rent, complete with servants who will take care of every need. Many holidaymakers from India make their way here.

flight takes about 35 hours. There is a lot to see, since the area has been inhabited continuously for many thousands of years. It is easy to travel, even though there are few cars. Train service is fast, reliable, and cheap; it only costs about 4¢/km. There are some disadvantages to travel in India, though. Sharks are a constant threat to sea-bathers. Drinking water and milk should be boiled, and only fruit that has a peel or rind that can be removed should be eaten. (Why?) If you are a meat-eater, stay away from southern India, because the people there are mainly vegetarians! Of course, you are on holiday, and you might enjoy *masala dosa*, a huge, crisp pancake stuffed with vegetables and served with yoghurt curds.

Industries in Pakistan are rather more down-to-earth when compared with India's young tourist industry. Pakistan is very short of steel (absolutely essential for industrialization) and just completed a new steel plant agreement in 1974. The plant is to be located at Pipri, about 40 km south of the old capital, Karachi. The Soviet Union sent 700 technical advisers to help build the plant and 1200 Pakistanis were sent to the U.S.S.R. for technical training. Many of the Soviet personnel involved in this steel mill have also had experience with steel plants in India. When the plant is opened, in 1978, Pakistan will probably save more than $100 million on iron and steel products, which currently have to be imported from the U.S.A. and Japan.

Bangladesh is short of just about everything except

Defenders of the Khyber Pass for centuries, the Pathans or Pushtuns of Pakistan would like to split away from Pakistan and form their own state.

people. After a few years of independence, the loss of mainly Pakistani managerial talent following the war of independence has made itself felt. Prices soared 200% to 300% between 1971 and 1973. Industrial output dropped by more than 30% over the same time period. Jute, a vegetable fibre and Bangladesh's biggest earner of foreign exchange, was reduced in output by 28% over the same period. The biggest problem to be faced is overpopulation. If the population continues to grow at today's rate, it will double to more than 150 000 000 people by the end of the century. How can they be fed in a country where men are lucky to find work that pays $1 a day?

Nepal's chief claim to fame is as the home of the Gurkhas, a tribe of warriors who have served for over 100 years in the British Army. Ruled by a king, the Nepalese are fiercely independent but recently had to give in for a while to Chinese pressure over Mt. Everest. Mountain-climbing expeditions were curtailed so long as the Chinese complained about possible spying activities. The capital of the country is Katmandu, nowadays a tourist mecca. China and the U.S.A. compete for second place in the foreign aid race. First place is occupied by India. The advantages of foreign aid for Nepal are obvious to visitors. Modern roads and a railway, airstrips and telecommunications have "opened up" the land.

Bhutan and Sikkim are very small nations, indeed. With very little industry and small numbers of people, their hopes of remaining independent are not very bright. Their chief importance is as "buffer" states between India and China.

Sri Lanka, formerly Ceylon, has been referred to as the Indian earring, lying as it does at the southern end of the peninsula. It is governed democratically. The major exports of the country are tea, rubber, and coconut products. While the population is mostly Sri Lankan, there are perhaps 500 000 Indian residents who are all likely to be repatriated to India under the terms of the Indo-Ceylonese agreement of 1964. In addition, there are 350 000 Indians who have taken out Sri Lankan citizenship. There is a danger that, when the non-citizens leave, the quality of work done on the tea and rubber estates will fall. It has been found through experience that Sri Lankans do not cope very well with rigid time schedules, yet rigid time schedules are vital when picking tea. The leaves must be picked exactly on schedule or else the quality is impaired. Tea accounts for nearly 70% of Sri Lanka's foreign ex-

The Nepalese *Gurkha* is reputed the most fearless warrior in the world, armed as he is with a *kukri*, a short, heavy, curved knife which may not be re-sheathed without drawing blood.

The ruler of Nepal is also a king. Birendra Bir Bikram Shah Deva was educated at Eton, Tokyo University, and Harvard. He is a modernist and is suspected of sneaking out of his palace to travel incognito on his motorcycle.

The king of Sikkim is known as the *Chogyal* and is venerated as a god. The Sikkimese are outnumbered 3 to 1 by immigrant Nepalese and seem fearful of being annexed by India because their country lies along the Chumba Valley. This valley is of great importance for the defence of India because it is a convenient route for invasion from the north.

The first tracks of the abominable snowman were seen in Sikkim in 1887.

The women who pick the leaves off tea bushes in Sri Lanka are called "plunkers". Plunkers earn 2¢ for every 0.5 kg of tea they pick. To earn 60¢ a day, they must work all the daylight hours.

change. You might wonder in that case why, when Sri Lanka became independent in 1948, the Indian tea estate workers who had come to work there under British rule were deprived of their rights to vote as well as their citizenship. It was due to the fears of a small nation that a large minority group such as the Indians would not fit in very well.

In the world of the jeweller, the mention of Sri Lanka brings a gleam to the eye. Oyster beds (genus *margaritifera*) on the north side of the island stretch for 2500 km². The last major exploitation of these pearl beds took place in 1956; there is still a major resource of natural pearls there. Furthermore, the Japanese are interested in opening up a pearl-culturing industry as a joint venture with Sri Lanka. In Japan and Australia it takes 5 to 6 years for cultured pearls to grow. In Sri Lanka it takes only 3 years, due to freedom from typhoons and tsunamis, violent marine upheavals.

Tea, rubber, and coconuts form the basis of Sri Lanka's foreign trade.

For tourists in Sri Lanka, the land produces sapphires that you can mine for yourself on payment of a fee of 56¢ to the government. In addition, you stand a good chance of finding rubies, moonstones, garnets, cats-eyes, or other precious and semi-precious stones.

Above all, though, the Indian sub-continent needs to produce things that will allow it to become self-sufficient in food. This is the major problem. Without enough nutritious food, people do not feel well. They become lethargic and prone to diseases, if they are not already suffering from diseases of deficiency in vitamins or minerals. It becomes impossible to plan effectively if one is always wondering where the next meal will come from, as well as not being able to work at peak capacity. The Indian sub-continent has 2 choices, then. It can either try to increase per capita food production, or it can try to tailor the number of people to fit the available food supply. In practice, a little bit of both is being done. The countries are trying to feed the people who are already there, as well as looking for ways to curb the natural increase of the population. Time is not on their side, as Figure 15-10 shows.

Well over 150 years ago, a scientist named Robert Malthus decided that population increased *geometrically*. That is, 2 people can produce 2 more, then these 4 can produce 4 more, these 8 can produce 8 more, and so on. In the meanwhile, food supplies increase only *arithmetically*. That is, food supplies go from a low level to a

Calories available per person per day in the form of:

Country	cereals	potatoes and starches	sugar	pulses and nuts	vege-tables	fruits	meat	eggs	fish	milk	oils and fats	total
Bangladesh						data not available						
Bhutan						data not available						
India	1 340	36	170	175	2	27	6	1	4	97	78	1 940
Jammu and Kashmir						data not available						
Nepal	1 722	42	12	40	7	12	15	4	2	91	86	2 030
Pakistan	1 691	35	197	56	13	48	20	1	6	156	125	2 350
Sikkim						data not available						
Sri Lanka	1 339	86	243	312	35	14	8	8	37	36	88	2 210
Canada	646	147	529	61	68	111	672	58	22	380	457	3 150

Figure 15-10.   Eating Habits of the Indian Subcontinent Compared with Those of Canada. The SNU or Standard Nutrition Unit requires that the population of a country have enough food so that on average **each member of a country's population has 2464 Calories available per day.**

higher level only with great difficulty. The result is as follows:

population   2 ──▶4 ──▶8──▶16
food   1 ──▶2 ──▶3 ── 4   riots, starvation.

Although Malthus was writing about Europe, could he have been prophesying for the Indian sub-continent? What do *you* think?

QUESTIONS AND EXERCISES

1. Construct a graph to show the density of population for each of the countries mentioned in Figure 15-3.
2. What is the average size of a farm in India? In all the other sub-continental countries? In Canada? Graph your answers. Find out what the word *gavelkind* means. How does it apply to Indian agriculture?
3. Devise a way of showing the impact of the Green Revolution in the countries mentioned in Figure 15-8. In addition, show the various natural and man-made disasters that affected production of the rice crop in the sub-continent. What are *hybrid* seeds?
4. For the 4 countries listed in Figure 15-9, work out a way of grading them according to "the quality of life."
5. Discuss the information shown in Figure 15-10, and write down why Canada's diet seems to be so much different from the diets of people in the Indian sub-continent. What is *kwashiorkor*? Which country do you think is likely to have most kwashiorkor? Find out as much as you can about *deficiency diseases*. Write short notes on the ones likely to appear in the Indian sub-continent.
6. Describe some of the problems likely to face the ruler of any Indian sub-continental country if he or she tried to computerize income-tax returns.
7. Is it true to say that the monsoon rules India?

16. THE U.S.S.R.

AREAL DIFFERENTIATION

The Union of Soviet Socialist Republics is, in terms of land area, the largest country in the world. In Arctic latitudes, for example, it stretches nearly halfway around the world. This means, of course, that as a Murmansker is yawning and stretching, knuckling the sleep from his eyes and thinking of the day ahead, a Siberian miner is probably just sitting down to supper. In a north-south direction, too, the U.S.S.R. is immense. From the icy wastes of the frozen Arctic Ocean the land stretches southwards to another desert — the great temperate desert of south-central Asia. It is truly a vast land.

The Russian Soviet Federal Socialist Republic is the largest of the republics, not only in land area, but also in population. In fact, the R.S.F.S.R. occupies 76% of the total land area of the U.S.S.R. and contains approximately half the population of the U.S.S.R. Not surprisingly then, this republic also contains the U.S.S.R.'s capital city, Moscow.

The whole U.S.S.R. occupies 16.7% of the world's land area, yet only 6.6% of the world's people live there. If we take the second figure and divide it by the first, we have a new figure which we can call the *population density coefficient*. For the U.S.S.R. it is 0.4, for Canada it is 0.09 $(\frac{6.9}{9.0})$, while for the world as a whole, including the U.S.S.R. and Canada, the PDC is 1.0 $(\frac{100}{100})$. Countries such as Japan have much greater values for their PDCs because the people are more crowded together. Japan's value is 10.0 $(\frac{2.8}{0.28})$. So you can see that it is 10 times more crowded than the rest of the world but rather less crowded than the Netherlands, which has a value of 12.0 $(\frac{0.36}{0.03})$.

A figure such as the population density coefficient can only be used to compare one country with another, and with the world average. It actually tells very little about where most people tend to live inside a given country. In the case of the U.S.S.R., we can overcome this difficulty quite well if we consider the actual populations of the republics shown in the map for Figure 16-1.

It is possible to calculate population density coefficients

The U.S.S.R. is so large that it takes more than a week for a train on the Trans-Siberian railroad to go from Moscow to Vladivostock, a distance of 9280 km.

The U.S.S.R. is equivalent in size to 252 Portugals or 70 006 Maltas or 3 Australias or 2 Canadas or 2044 Jamaicas or 9 Algerias.

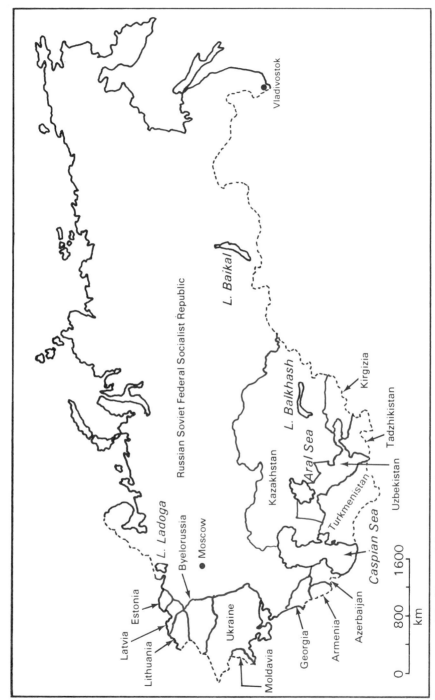

Figure 16-1. The Union of Soviet Socialist Republics

	Population 1973 (estimated)	% increase since 1966	Land area square kilometres
Russian Soviet Federated Socialist Republic	132 189 000	3.9	16 701 000
Ukraine	48 237 000	5.9	594 000
Byelorussia	9 202 000	6.3	205 000
Moldavia	3 722 000	10.5	34 000
Estonia	1 405 000	8.3	45 000
Latvia	2 430 000	6.6	63 000
Lithuania	3 233 000	8.2	51 000
Armenia	2 667 000	19.1	30 000
Azerbaijan	5 421 000	16.8	87 000
Georgia	4 835 000	7.7	71 000
Kazakhstan	13 695 000	13.7	2 721 000
Kirgizia	3 145 000	20.3	196 000
Tadzhikistan	3 188 000	24.7	140 000
Turkmenistan	2 360 000	23.1	482 000
Uzbekistan	12 896 000	24.7	394 000
USSR (total)	248 625 000	7.1	21 814 000

Note: Land areas exclude water surfaces.

Figure 16-2.   The Republics of the U.S.S.R.

for each of the republics and show them on a map. However, such a map would not accurately reflect the true distribution of population within the U.S.S.R. A much better picture of population distribution is provided by the map shown in Figure 16-3.

Figure 16-3 ignores man-made boundaries. As you can see, the population of the U.S.S.R. is concentrated into a relatively small part of the land area. There are, of course, good reasons for this. People anywhere will only be content to settle where they are likely to make the best living that they can. Why else did you or your ancestors come to Canada? And why did you or they settle in that particular part of Canada? People in the U.S.S.R. are no exception to this rule. If you look at Figure 16-4, you will see that the inhabited land area is actually quite a small part of the U.S.S.R.

Cities and towns take up less than 0.1% of the land area of the U.S.S.R. But since the great majority of people (and therefore cities and towns) are situated in the farming regions (how else would they be fed?), then clearly, the *inhabited portion* of the U.S.S.R. is less than 30%. This pushes the PDC for the *inhabited portions* of the U.S.S.R. up to 1.3. In other words, the presently habitable part of the U.S.S.R. is more crowded than the

The cities of the U.S.S.R. have been subject to exactly the same problems of growth and pollution that Western countries have been experiencing recently. With one exception: the U.S.S.R. is only just entering the age of the automobile. Cars are very expensive in the U.S.-S.R. It takes several years' salary for ordinary people to buy one, and the order must be placed months or years in advance.

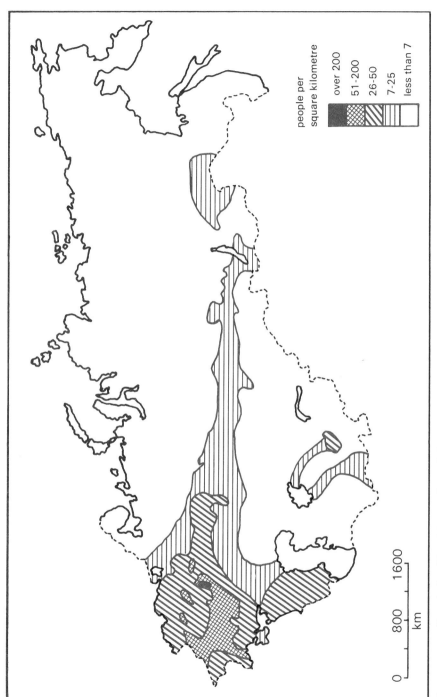

Figure 16-3. Population Densities in the U.S.S.R.

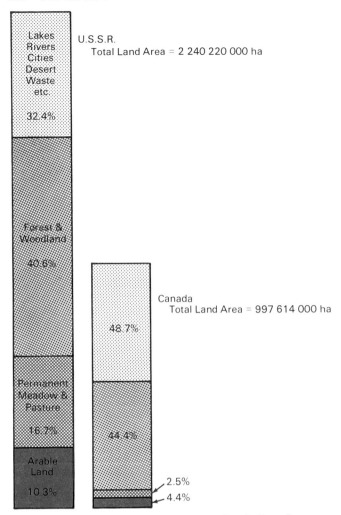

Figure 16-4.   Soviet Land Use Compared with Canada

world as a whole. In geographic studies, though, a simple statement like that is insufficient. Why does the population not spread itself out a lot more? Has it ever tried to? Is it trying now? Are there plans to do so in the future?

Before we try to answer those questions, let's examine some of the physical influences at work on the landscape of the U.S.S.R. that prevent the population from spreading more.

PRECIPITATION

Annual totals are not especially high anywhere in the Soviet Union. Indeed, there are 2 great deserts in the

Unmarried people pay a higher rate of tax than married people in the U.S.S.R. The country desperately wants a larger population to fill in the eastern empty spaces.

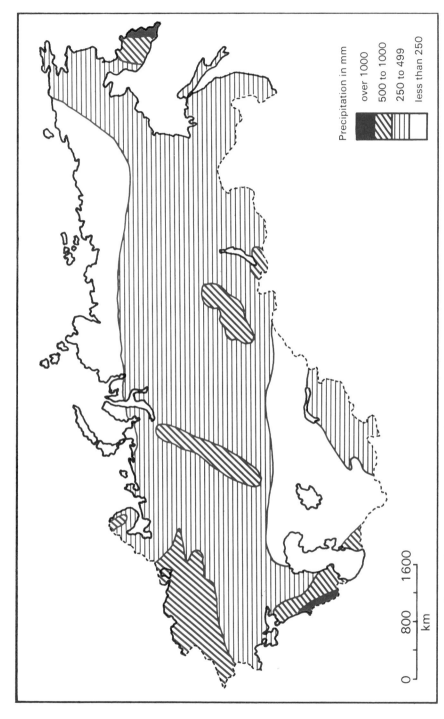

Figure 16-5. Total Annual Precipitation in Rainfall Equivalent in the U.S.S.R.

U.S.S.R., one in the Arctic and one in the south. In addition, much of the moisture precipitated in these two areas is lost to plants because either it is frozen (as in the northern desert, the *tundra*), or it is quickly lost by evaporation (as in the southern desert).

TEMPERATURES

As is clearly visible from the maps in Figure 16-6, the U.S.S.R. is bounded on nearly all coasts by frozen ocean in winter. This has been the reason for attempts in the past century to open up the Black Sea to Russian shipping, to gain territory on the Baltic Sea, and to extend the eastern boundary of the R.S.F.S.R. south to escape the pack and drift ice and so keep Vladivostok open as a port all year 'round.

Central Siberia has very cold winters and hot summers. The extremes of temperature are as great as anywhere in the world. One consequence of this is that there is a widespread region of permafrost, where permanent frost invades the ground. In some areas, the permafrost melts in the top 25 cm or so and shallow lakes and streams are to be seen during the short tundra summer of perhaps 2 months. Only mosses, coarse grasses, and a very few stunted shrubs and trees will grow. No farming can be performed on the permafrost.

The growing season is completely dependent upon the annual range of temperature. Generally, the growing season anywhere needs to be at least 5 months in length — that is, at least 5 months in the year should have average temperatures in excess of 6°C. If the growing season is too short, then farming will not be possible, for crops will not grow and ripen properly. 6°C is the temperature below which most plants stop growing and lie dormant.

NATURAL VEGETATION

If people were not in existence and so had no influence on the land surface at all, the only plants that would grow would be *natural vegetation*. Far to the north and pretty well inside the Arctic Circle lies the *tundra*. Here mosses, lichens, coarse grasses, and some stunted trees are all that are likely to grow. The long Arctic winter poses further problems for plants, in that the farther north one goes, the length of time that the sun never rises increases. On the Arctic Circle itself, the 21st of December has 24 hours

Perfectly frozen mammoths have been found in Siberian permafrost. Although thousands of years old, the meat is edible, often in top condition, and has been tried by humans who have suffered no ill effects.

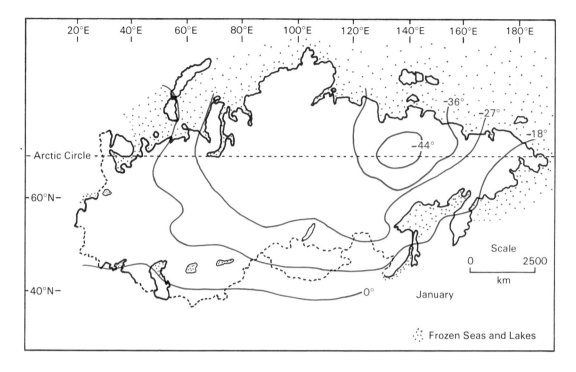

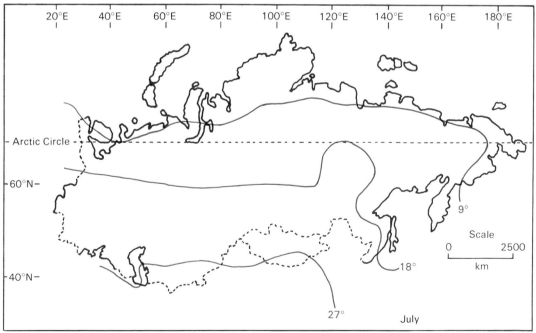

Figure 16-6. Average Temperatures for January and July in the U.S.S.R. (° Celsius)

of night. At the North Pole, the period of darkness has expanded to fill 6 months, from September 21 to March 23. Of course, this is balanced by a summer of 6 months, too. It is possible to grow crops under artificial conditions so that they profit from the long summer days of the northern U.S.S.R., but otherwise soils are too poor, and precipitation and temperature are too low.

South of the tundra is the *taiga*. This great coniferous forest covers a vast area and is relatively little exploited. To the west of the Yenisei River is the West Siberian Forest of Siberian fir, stone pine, spruce, silver fir, and Siberian larch. East of the Yenisei River is the Eastern Siberian forest of Siberian fir, stone pine, and eastern larch. The southeast Siberian transitional forest contains a wide variety of tree types: eastern larch, Siberian fir, ayan pitch-pine, Manchu pine, oaks, elms, maples, walnuts, and wild apples. Right over on the western side of the Soviet Union is to be found the eastward limit of the Fenno-Scandian forest, with pines, spruce, and birch.

The taiga thins out southwards through a zone called the *wooded steppe* and passes gradually into grasslands known as *steppe*. Here summer temperatures increase, so that any summer rainfall nearly all evaporates. Minerals and organic plant debris stay at or near the surface of the soil, thereby making it very fertile. Black in colour, the soil of the steppe is known as "Black Earth" or *chernozem*, which means the same thing in Russian. (Much of the Canadian prairies is steppe.) The Russian steppe is a zone of intense agricultural activity, and nearly all the original grassland has disappeared. In the European part of the U.S.S.R. and lying just to the north of the steppe is a tapering wedge of hardwood forest that once covered most of Europe. Oaks, beeches, maples, walnuts, chestnuts, mulberries, and the like typify this region, but here too the soil is very fertile, and the forest has largely been cut down.

At the latitude of the Caspian Sea begins the *salt steppe*. It is a semi-arid land; trees here are mostly acacias and tamarisks. Irrigation is needed to produce crops, but fresh water in this area is hard to come by, since the Caspian and Aral Seas are both salty. Then, finally, there is the *desert* proper. Here little grows save for cactus and other drought-resistant plants.

You can see that the U.S.S.R. is a vast area, with great changes in landscape. Yet there are really very few people to fill it. The few people that there are, are all crowded

The steppes have been the traditional route of invaders from the East into Europe. Genghis Khan, the Mongol, and Attila the Hun swept out of the limitless, grassy wastes with their nomadic, herding peoples. From the Ural Mountains to the English Channel there is nothing but an undulating plain, ideal, first, for cavalry and, later, mechanized warfare.

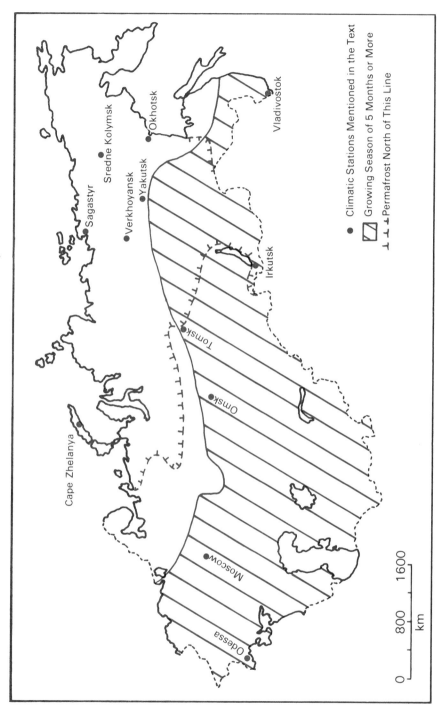

Figure 16-7. Two Climate Controls on the Agriculture of the U.S.S.R.

		J	F	M	A	M	J	J	A	S	O	N	D
Moscow	a	−9	−8	−4	4	13	17	19	17	11	4	−3	−8
	b	38	36	28	48	56	74	76	74	56	69	43	41
Omsk	a	−22	−18	−12	−1	9	16	18	16	11	1	−11	−18
	b	15	13	13	15	30	51	51	51	28	25	18	20
Odessa	a	−4	−2	2	9	15	20	23	22	17	11	5	−1
	b	25	25	25	25	38	51	51	38	38	25	38	38
Cape Zhelanya	a	−17	−16	−21	−17	−8	2	2	2	0	−4	−13	−17
	b	5	3	3	3	5	13	25	33	18	8	3	3
Tomsk	a	−21	−17	−11	−2	8	14	17	15	9	0	−12	−19
	b	28	18	20	23	41	69	66	66	41	51	46	38
Irkutsk	a	−18	−19	−11	−1	7	13	16	14	8	−1	−12	−20
	b	13	10	8	15	33	56	79	71	43	18	15	15
Verkhoyansk	a	−50	−44	−32	−13	2	13	16	11	2	−14	−37	−46
	b	5	2	2	5	7	22	24	24	12	10	7	2
Yakutsk	a	−45	−37	−23	−9	4	14	17	14	6	−8	−29	−41
	b	8	5	3	8	10	28	41	33	25	13	10	8
Sagastyr	a	−37	−38	−34	−22	−9	0	5	3	1	−14	−27	−33
	b	2	2	—	—	5	10	7	34	10	2	2	5
Sredne Kolymsk	a	−42	−37	−28	−17	−3	10	13	8	2	−12	−29	−38
	b	8	10	5	5	5	23	33	30	20	13	13	10
Okhotsk	a	−24	−22	−15	−10	1	6	11	12	−8	−3	−15	−22
	b	3	3	5	10	23	41	56	66	61	25	5	3
Vladivostok	a	−14	−10	−3	4	9	14	18	21	16	9	−1	−10
	b	8	10	18	30	53	74	84	119	110	48	30	15

Figure 16-8(a). Climatic Data for 12 Cities in the U.S.S.R.
a = average monthly temperatures °C
b = average monthly precipitation (in rainfall equivalent) mm

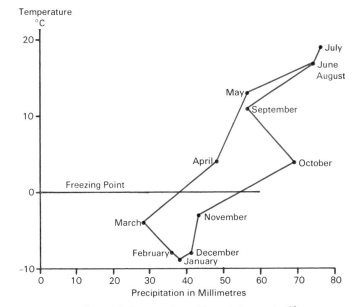

Figure 16-8 (b). Hythergraph to Show Moscow's Climate

into one major area. How does the U.S.S.R. hang together? How can it function as one country?

For several centuries the Russian empire expanded first east to the Pacific and then south to the mountain ranges of central Asia: the Caucasus, the Tien Shan, and the Altais. Under the czars, the Russian empire expanded to fill in political vacuums left by the overthrow of the Turkish and Chinese empires in the late 19th century. For almost all of that time, since 1300 when the Russian Empire was just a tiny nucleus of a state, centred on Moscow and ruled by the Princes of Muscovy, the main attention of the reigning emperor (czar) was concentrated on the European part of the country. Siberia, that vast portion of the R.S.F.S.R. which begins on the eastern side of the Ural Mountains, became a vast prison. Politically dangerous people and other persecuted unfortunates were shipped there in millions — exiled, never to return home. Sometimes they were simply confined to a given area. At other times, the prisoners were allocated as slave labour on projects such as the Trans-Siberian Railway, which had been constructed under the czars in the late 19th century in response to developments in railroad-building in Canada and the U.S.A. This was then the only link from coast to coast, unless we consider the road network. But roads in the U.S.S.R. were often only dirt, boggy and impassable in spring, hidden in winter, and dust bowls in summer. Before the development of air transportation, most parts of the Soviet Union were effectively isolated.

After the Bolshevik take-over following the revolution of 1917, the Soviet Union started on a new chapter in its history. Industrialization was stepped up to a massive scale. Farming was reorganized along collective lines in the hope of raising efficiency. Political turbulence and the access of Joseph Stalin to power resulted in the arrest and conviction of millions of people, who were allocated to slave-labour programs just as enemies of the czars had been. The steppe east of the Urals was ploughed for the first time in an attempt to produce more food. Vast new state farms were marked out and ploughing began. Natural resources of iron were tapped by one of the largest iron and steel plants in the world at Magnitogorsk in the southern Urals. Coal in the Kuzbas Basin, near the sources of the Ob and Irtysh Rivers, was used as a source of power for new cities, which were to produce a wide range of industrial products, including textiles. All this

The name of one of the Soviet newspapers is *Pravda*, which means "truth."

Novosti Press Agency

Oil fields in the U.S.S.R. today. Tyumen (above) is one of the latest to be discovered.

was made possible by the Trans-Siberian Railway. It formed the life-line of the newly expanding part of the Soviet economy.

In 1942, during the Second World War, the Soviet Union was attacked by Germany. Industrial areas in the west were quickly overrun, and a great deal of Soviet industry was transferred to the east behind the Urals, thereby boosting the area's industrial capacity. At the close of the 1940s, after the war, the U.S.S.R. was rebuilding the ravaged west but still depended on the east a great deal, especially for food. However, in the late 1950s and onwards, attention was increasingly paid to the natural resources of the Siberian northland. What a treasure house it proved to be! Coal, oil, natural gas, diamonds, copper, gold, iron ore, and many other minerals have been discovered there in vast quantities.

In order to exploit these riches the U.S.S.R. is contemplating another railway, designed to tie together vast areas of the land that are at present inaccessible. To be completed in the 1980s, it will stretch for 3200 km across Siberia, well to the north of the existing line. It will employ 70 000 workers directly in its construction, and will cross gorges, marshes, and fast-flowing rivers. Long tunnels will carry the double-tracked railway through high mountains. In order to encourage people to settle in Siberia, rather than force them as formerly, the Soviet government is paying resettlement grants and loans, giving tax exemptions for 8 years and generous leaves with pay. The U.S.S.R. is determined to populate its far east. Its southern neighbour there is teeming China, and China claims that a great deal of eastern Siberia was taken illegally by Russia during the 19th century. They would like to have it returned.

Not a great deal is known about the exact extent of the mineral belts of the U.S.S.R. The Soviet government is fairly secretive about it, but it is very likely that the U.S.S.R. has *at least* half the world's known coal reserves. In addition, oil and natural gas finds are adding to the proven reserves of the Soviet Union at a tremendous rate. In the 20 years from 1950 until 1970, proven oil and gas reserves increased by at least 5 times, and new discoveries continue to be made in the Siberian plain just east of the Ural Mountains. In 1973, the Tyumen region near the mouth of the Ob yielded 20 000 000 t of oil in the first three months. It is hoped that 600 to 700 million tonnes will be recovered annually from this one field alone. Al-

Women in the U.S.S.R. are a large part of the work-force. An official death toll of 20 000 000 people in the Second World War gave women a working equality unequalled in the West. They outnumber men 3-1 in the medical and teaching professions. Every third engineer or judge is a woman.

In the late 1960s, Otari Lazishvili was sentenced to 15 years in jail for setting up an underground factory network to produce consumer goods such as nylons and shopping bags that were not readily available.

In the effort to populate the region of Siberia, the Soviets rely on encouraging people — especially young people — to go east. In northern Siberia, for example, the pay is 70% higher than in Moscow. Welders, pipelayers, and foremen earned $930 to $1060 a month in 1973, and a truck driver managed $530 a month — a good wage when you consider that the average Soviet monthly wage was only $170 in that year!

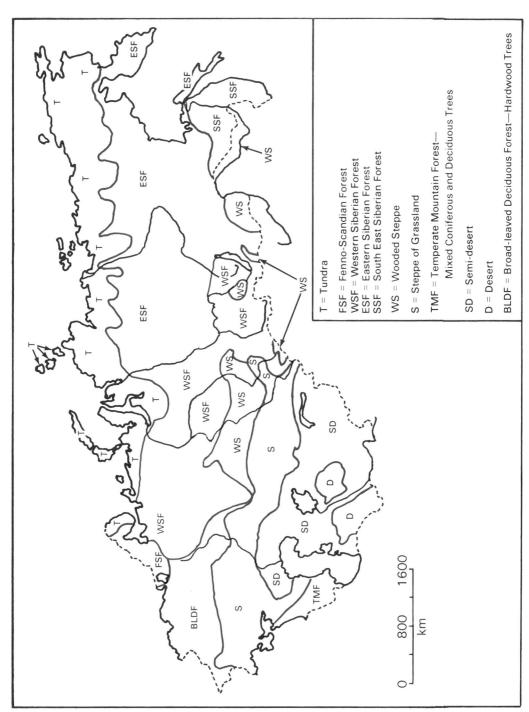

Figure 16-9. The Natural Vegetation of the U.S.S.R.

T = Tundra

FSF = Fenno-Scandian Forest
WSF = Western Siberian Forest
ESF = Eastern Siberian Forest
SSF = South East Siberian Forest

WS = Wooded Steppe

S = Steppe of Grassland

TMF = Temperate Mountain Forest—
        Mixed Coniferous and Deciduous Trees

SD = Semi-desert

D = Desert

BLDF = Broad-leaved Deciduous Forest—Hardwood Trees

0   800   1600
        km

ready 2900 km of oil pipeline and 1900 km of gas pipeline have been laid. Such large investments are difficult, if not impossible, for the U.S.S.R. to finance by itself. Thus foreign capital is being sought, chiefly from the U.S.A. and Japan. The Japanese will have to invest at least $1 billion in order to build a pipeline from the Pacific to Irkutsk — a distance of 4000 km. The Soviets themselves have already connected Irkutsk with Tyumen by means of a 2500 km pipeline. The U.S.A.'s giant power companies such as Gulf, Occidental Petroleum, El Paso Natural Gas, and Cyrus Eaton's industrial complex are contemplating spending $10 billion to $15 billion in order to run a gas pipeline to Murmansk on the White Sea. Here special facilities would liquefy the natural gas and pump it aboard specially constructed vessels for the voyage to the U.S.A. It is expected that both schemes will be in operation by 1980 at the latest. It is no wonder, then, that the Western world looks on the Soviet development of Siberia as "another Klondike, a new gold rush, a bonanza." Certainly the Soviets themselves think along these lines. It is their new frontier. Not only the financial incentives lure people away from the western U.S.S.R. but also the feeling of rawness, newness and wilderness so attractive to many people the world over. In addition, the builders of the new cities of the north and east feel that they are in on the start of something big, something to be proud of. East of the Urals is the frontier, the challenge and the hope of great reward, both personal and national.

Agricultural development east of the Urals is also a top priority, and has been for decades. The greatest increase in Soviet farmland took place in the years from 1950 to 1963, when the total sown area in the U.S.S.R. increased by 50%. Most of this development took place in the *New Lands*, between the Volga and the Yenisei Rivers. Great hopes were placed on this area for wheat production so that it would more than take up the slack as much of the Ukraine, long the wheat centre of the U.S.S.R., changed over to other crops, mainly fodder. This was designed to boost the supply of meat for the Soviet people, most of whom live in European Russia, as we have already noted.

In fact, the U.S.S.R. has had great problems with the New Lands. Wheat yields are about 50% lower than on the Ukraine's famed black earth, largely because wheat is pretty well grown exclusively (monoculture). In addition fertilizer is in short supply, and rainfall is often a

A new settler in Siberia has travel expenses paid and receives a grant of 4 months' salary. In addition, he or she receives 36 working days of holiday annually, 12 more than normal elsewhere in the U.S.S.R.

In June 1974, Pepsi-Cola came to the U.S.S.R. It sold at 55¢ a bottle. Coca-Cola signed a deal with the Soviets that month, too, but not only to produce beverages. They planned to cooperate in such things as growing fruit and vegetables on waste land, purifying and desalinating seawater, processing tea products, and developing enriched drinks from milk wastes.

Finland has thousands of its nationals building a steel factory in Karelia to take advantage of local iron ore deposits at Kostamos. Annual production will total 30 000 000 t of ore out of a total deposit of 1.2 billion tonnes. This fits in well with a shortage of Soviet manpower and capital, and a shortage of Finnish jobs and iron ore. In addition, the Finns are helping to build a gigantic forest-products factory in Svetogorsk, Karelia.

To boost grain production in 1973, the Soviet government offered cash prizes of up to $180 000 for the top producers.

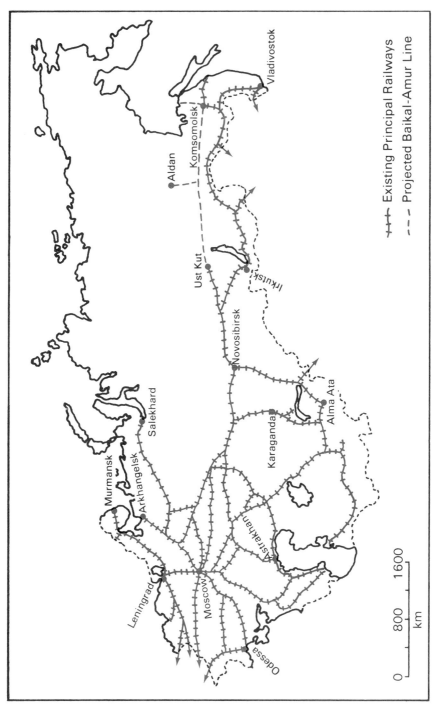

Figure 16-10. Principal Railways of the U.S.S.R.

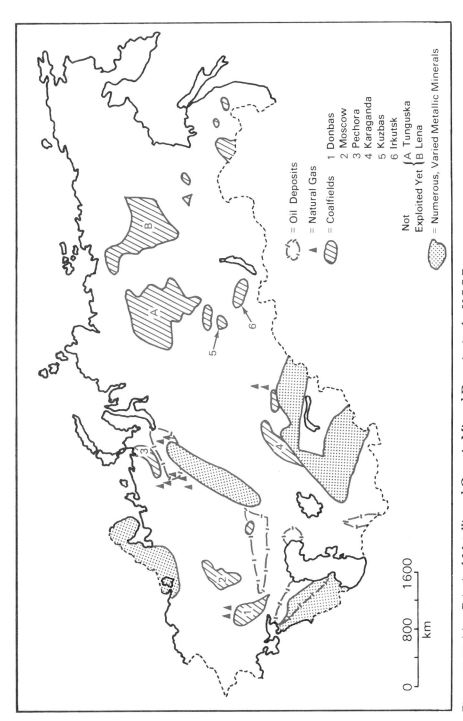

Figure 16-11 (a).   Principal Metallic and Organic Mineral Deposits in the U.S.S.R.

Product		Output USSR	Output by largest world producer		Output — World Total	Units of measurement
Coal	Reserves	4 121 603	USSR		6 641 200	million tonnes
	Production	441 416	USA	503 050	2 124 000	million tonnes
Crude	Reserves	8 203 000	Saudi Arabia	18 737 000	76 200 000	thousand tonnes
Petroleum	Production	377 075	USA	466 704	2 399 400	thousand tonnes
Natural	Reserves	18 010 000	USSR		49 900 000	million cubic metres
Gas	Production	212 398	USA	636 931	1 142 400	million cubic metres
Iron Ore (iron content)		110 341	USSR		424 792	thousand tonnes
Bauxite		4 400	Jamaica	12 543	625 000	thousand tonnes
Chromium		765	USSR		2 980	thousand tonnes
Copper Ore (copper content)		990	USA	1 381	6 390	thousand tonnes
Lead		460	USA	525	3 400	thousand tonnes
Manganese		2 552	South Africa	1 368	8 100	thousand tonnes
Mercury		1 724	USSR		8 870	tonnes
Nickel		118 000	Canada	266 664	681 000	tonnes
Silver		1 213	Canada	1 398	9 096	tonnes
Tungsten		8 800	China	101 000	45 440	tonnes
Zinc		610	Canada	1 268	5 390	thousand tonnes
Salt		11 968	USA	39 986	139 313	thousand tonnes
Asbestos		1 152	Canada	1 489	4 851	thousand tonnes
Diamonds		8 800	Zaire	12 677	46 358	thousand metric carats
Installed Electrical Capacity	total	175 365	USA	386 701	not available	thousand kilowatt
	(hydro)	(133 448)	(USSR)		not available	
Electrical Production		800 400	USA	1 717 521	5 222 500	thousand kilowatt hours
Energy Consumption		1 112	USA	2 327	7 088	million tonnes of coal equivalent
Nuclear Energy Production		2 031	USA	8 687	25 020	thousand kilowatt hours
Steel		120 660	USSR		575 061	thousand tonnes
Radios		8 794	Japan	28 091	80 370	thousands
Televisions		5 814	Japan	13 231	44 764	thousands
Motor Vehicles	cars	529	USA	8505 (2053 trucks)	25 361	thousands
	trucks	862	Japan	2105 (3718 cars)	7 110	thousands
Telephones		4.9	USA	60.4	7.8	per hundred people

Figure 16-11(b).   Soviet Mineral and Industrial Production Compared with the Rest of the World, 1971

problem. Wheat yields in the New Lands are actually diminishing, even though an additional 270 000 km² have been planted with wheat since 1950. To make up the deficit, made even more acute by the increase in population encouraged by the government, the U.S.S.R. has resorted to wheat purchases abroad, chiefly from the U.S.A. and from Canada. After the poor harvest of 1963, the Soviets had to buy 12 000 000 t of grain from North America. In 1970 and 1971, they bought 3 300 000 t of grain. In 1972, the Soviet grain deficit was about 25 000 000 t, and in the world's largest grain deal ever, they bought $2 billion worth of grain from Australia, Canada, the U.S.A., France, and Sweden — altogether about 28 000 000 t of it.

The Soviet Union has been trying for a long time to boost meat production. In 1971 the government planned to build 1170 new mechanized state farms to produce red meat and 585 poultry farms, *all within 3 years*. But they are short of fodder crops and feed grain for the animals to eat.

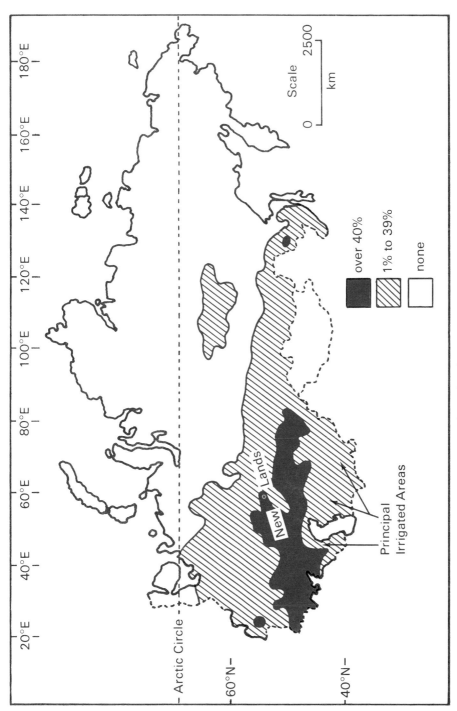

Figure 16-12. Soviet Arable Land as a Percentage of Total Cultivated Land

	1961		1962	1963	1964	1965	1966	1967	1968	1969	1970	1971	1972
Wheat Production thousand tonnes	66 483		70 778	49 688	74 399	59 686	100 499	77 419	93 393	79 917	99 734	98 760	85 8C
Population thousands	218 150		221 730	225 060	228 150	230 940	233 530	235 990	238 320	240 550	242 768	245 066	247 34

Figure 16-13.   Wheat Production and Population Increase in the U.S.S.R.

It is fair to place part of the blame for such poor performances by the agricultural industry on farm organization. Collective farms known as *kolkhoz* are found mainly to the west of the Urals. They occupy 48% of the U.S.S.R.'s farmland and account for about 52% of production. There are 37 000 collective farms, which were produced by combining individual landholdings in villages, and their average size is about 6235 ha. State farms known as *sovkhoz* are run just like any other state organization. They also occupy 48% of the U.S.S.R.'s farmland but only account for 15% of total production. There are perhaps 70 000 such farms, and their average size is about 4939 ha. The remaining 33% of agricultural production is produced on only 4% of the farmland — private plots averaging 0.2 ha in size, tended by collective or state farm workers in their spare time. But they are allowed to sell any surplus that they have in the nearest town or city. In this way, the U.S.S.R. has an assured supply of vegetables, at least.

Additional blame can be attached to the climate of the European part of the U.S.S.R. Drought in 1972 produced the disastrous grain harvest of that year, and the shortage of snow in the winter of 1972 and 1973 imperilled the 1973 crop. Snow is needed to provide meltwater in the spring to get the newly sown wheat off to a good start. Luckily, enough snow fell in time and the 1973 harvest was a record. To alleviate future drought, at least in the New Lands, a 450 km irrigation canal is planned from the Volga to the Ural River. Work started in 1974. But more than irrigation canals are needed. Trained personnel are also badly required for fertilizer plants, which often operate at 60% of capacity. In the 1960s, 72 new fertilizer plants were planned, only 27 of which were on schedule by December 1972. At that time, too, a lack of rail cars resulted in fertilizer piling up in the factories.

It is the sheer scale of the U.S.S.R. that poses the great-

Novosti Press Agency

The "Volga-Don" state farm was awarded the Order of Lenin because of its successes.

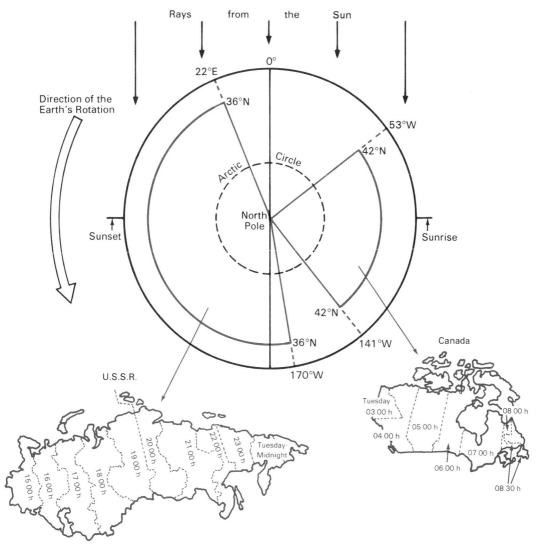

Figure 16-14.   The Latitudinal and Longitudinal Extents of Canada Compared with the U.S.S.R. The inset maps show their consequent, respective time zones. International Date Line omitted. The day shown is Tuesday, July 2, 1974. The Earth is not shown tilted.

est problems for its development. As we have already noted, it stretches about halfway around the world. You can compare this with Canada's extent in Figure 16-14.

The time zones on the continents are put in for our convenience, of course. A moment's thought should convince you that nowhere can different longitudes have the same time. One degree of longitude represents 4 min of time on a clock, 1 min of longitude represents 4 s, and 1 s of longitude represents 0.066 s on the clock. As can be seen easily from Figure 16-14, there is a great *longi-*

In summer the population of Moscow drops markedly as people move out to *dachas*, summer cottages in the countryside. Even poor people try to rent a couple of rooms in a quiet village.

tudinal extent to the U.S.S.R., and therefore a wide variation in time. To reduce difficulties and problems associated with this phenomenon, the U.S.S.R. is divided, like everywhere else in the world, into time zones. Everywhere within a zone has the same time — even though *really* it doesn't!

In order to conclude our study of the U.S.S.R. and obtain some kind of an overview of the different landscapes there, it is a good idea to examine some of the landscapes that the Soviet people see themselves.

The two major landscapes are those of the city and those of the countryside. Usually, the city is called an urban landscape while the countryside is known as a rural landscape. Just like all other industrialized countries in the world, the U.S.S.R. has the majority of its people living in cities. As industry has broadened its scope, producing more and more different products, the number of people needed to work in factories and offices has increased. Obviously, cities contain large pools of labour from which to draw workers, so that cities attract even more industries. And, on the other side of the coin, people are attracted to the places where they are likely to get jobs — cities. Just like the cities of the rest of the world, the cities of the U.S.S.R. are growing, and growing rapidly.

The U.S.S.R. is a land of truly varied landscapes, ranging from cities to farms, forests to deserts and lands of ice and snow. In all these widely differing habitats, the Soviet peoples contrive to make their livings in ways that change from one year to the next. Thus the landscapes of the U.S.S.R. are not static but are ever-changing, reflecting the needs and feelings of the people. In this respect, the U.S.S.R. is identical with all the other nations of the world. Although all people in the world try to adapt their surroundings to suit themselves, the people of the U.S.S.R. seem to do at least as much as and probably more than most in this respect.

Soviet industry is plagued with problems. Computers made in 1973 were at about the level of Western ones made in 1965. Industrial quotas are set and often reached only at the expense of quality and choice. For example, hats may be produced in only 2 sizes, or pants in only one colour. Many Muscovites find their *avoska* indispensable. The word means "just in case" and applies to a string shopping bag that is easily carried. If a store receives a consignment of something interesting or necessary, then out comes the *avoska* for the purchase. Light bulbs or zippers might be all sold out tomorrow, so buy some now, *avoska*!

In 1972, $5.7 billion worth of poor-quality or unattractive clothes remained unsold in the Soviet Union.

	1913	1926	1940	1950	1956	1960	1965	1970	1975	1980	1985
Total Population	159	147	192	180	200	216	231	243	255	271	287
Urban	28	26	61	71	87	106	123	139	155	174	194
Rural	131	121	131	109	113	110	108	104	100	97	93

Figure 16-15.   **Estimates of Urban and Rural Population (Millions) in the U.S.S.R.**

These men, all centenarians, are natives of Georgia, a republic famous for the longevity of its people.

Novosti Press Agency

QUESTIONS AND EXERCISES

1. Draw a map of the U.S.S.R. to show population density coefficients for each of the republics related to the U.S.S.R. as a whole.
    Example:

    $$\text{PDC of the U.S.S.R.} = \frac{100\% \text{ of the population}}{100\% \text{ of the land area}}$$

    $$= 1.0$$

    $$\text{PDC of the R.S.F.S.R.} = \frac{53.2\% \text{ of the population}}{76.6\% \text{ of the land area}}$$

    $$= 0.69$$

    This result means that the R.S.F.S.R. is rather less crowded than the U.S.S.R. as a whole.

2. Draw hythergraph shapes to compare Moscow (illustrated in Figure 16-8) with any other 4 widely differing climate types in the U.S.S.R. Be sure to have all your shapes on the same hythergraph blank.

3. In any way that you wish, construct graphs or charts to show the world standing of the U.S.S.R. in (a) mineral production and (b) industrial production.

4. Do some library research to discover what difficulties faced the U.S.S.R. when it tried to become an industrial nation in the late 1920s and the 1930s.

5. Write an essay outlining the difficulties you face in your job as a jet pilot on a regular schedule between Moscow and Vladivostok.

6. Explain the *trends* (it might help you to draw a graph) of the data shown in Figure 16-14.

7. Obtain a map of Canada and a map of the U.S.S.R. You might have them supplied to you by your teacher or you might decide to trace them yourself from an atlas. *Make sure that they are both to the same scale.* Carefully, mark on the capital cities. Then, in addition, mark on the positions of 6 major cities in each country and measure the distances between them and their capital by means of the scale.

8. Compare Canada with the U.S.S.R. under following headings: *land area, distances, habitable land, population, natural vegetation zones, minerals, farming.*

17. CHINA

CHANGE

The traditional Chinese way of life began to develop over 4000 years ago in the valley of the Wei River, a tributary of the Hwang (Yellow) River in north-central China. Find it in your atlas. It was not until nearly 2000 years had passed, however, that the people became organized into a recognizable nation. At that time (221-206 B.C.) the Chin dynasty did much to create a united state, even to the extent of possibly giving the Han people (for that is what the Chinese people are correctly called) the name of their country, *China*. The Chin emperors extended the Han way of life out to the delta lands of the Hwang and also southwards towards the mouth of the Yangtze. Then they took steps to protect this way of life from marauding attack by bands of northern (Mongolian) nomads by building the Great Wall. The work of the Chin emperors was continued and consolidated by the emperors of the Han dynasty, which held power from 206 B.C. to A.D. 220. The Han dynasty pushed Han control all the way through the mountains of southern China as far as the South China Sea, and it even sent out explorers to the western mountains and deserts. China was now virtually complete in its territory, and settled in its way of life.

Chinese dynasties and dates

Shang	1523 B.C. to 1027 B.C.
Chou	1027 B.C. to 221 B.C.
Chin	221 B.C. to 206 B.C.
Han	206 B.C. to 220 A.D.
Three Kingdoms	220 A.D. to 265 A.D.
Tsin	265 A.D. to 420 A.D.
Northern and Southern Kingdoms	420 A.D. to 618 A.D.
Tang	618 A.D. to 906 A.D.
Sung	960 A.D. to 1279 A.D.
Yuan (Mongol)	1279 A.D. to 1368 A.D.
Ming	1368 A.D. to 1644 A.D.
Ching (Manchu)	1644 A.D. to 1911 A.D.
Republic	1911 A.D. to 1948 A.D.
People's Republic	1948 A.D. to ?

Sue Mason

Part of the Great Wall of China.

Dynasties and capital cities of China

Shang	Anyang
Chou	Sian
Chou	Loyang
Chin	Sian
Han	Sian
Han	Loyang
Three Kingdoms	Nanking/Chengtu/Loyang
Tsin	Nanking
Northern and Southern Kingdoms	Nanking/Tatung/Sian
Tang	Sian
Tang	Loyang
Sung	Kaifeng
Sung	Hangchow
Yuan (Mongol)	Peking
Ming	Nanking
Ming	Peking
Ching (Manchu)	Peking
Republic	Chungking
People's Republic	Peking

At this same time in Europe, the Romans held sway over an empire of similar scope. We all know what happened to the Roman empire. It fell. But it was a slow crumbling, rather than a dramatic collapse, occurring right through the 300s and 400s. Eventually all that remained of it was a little area around Constantinople (also called Byzantium, and now Istanbul). By then the Roman empire could no longer keep back the invading barbarians ("bearded ones") who came from eastern Europe and central Asia. The Dark Ages of Europe had begun. In China, on the other hand, the Great Wall proved to be fairly (not completely) effective against the "bearded ones", and the Chinese empire survived. Another reason for its survival was that, whereas the Roman empire was a miracle of organization and military power the Chinese empire was simply a coherent *way of life*. The traditions of centuries can survive the occasional passage of armies and the occasional depredations of marauding bands.

The Chinese empire, and its way of life, continued to survive right through the numerous centuries after the Han dynasty. Dynasty succeeded dynasty, just as the Han had succeeded the Chin, and the people continued to live as their forefathers had lived. The only changes they noticed were in the rhythm of the seasons and the passing of time as they aged. To a certain extent this *sameness*

A village candy-maker in the 1930s.

Grace McBride

Grace McBride

A town worker in the 1930s.

Louis Tsao

No change over three generations.

of their existence can be attributed to the fact that the Chinese people were isolated from the changes that were occurring in Europe. As the people of Europe experienced the Dark Ages, the Chinese continued to live as their forefathers did. As the people of Europe experienced the Renaissance, the Chinese lived as their forefathers did. Not even contact with the West through Marco Polo made them curious about other ways. As the people of Europe passed through the Industrial Revolution, still the Chinese lived as their forefathers did. Cut off from Europe by distance, mountains, and deserts, and content in their own feeling of superiority in any event, the Chinese did not seek the stimulation of contact with another culture. They certainly did not welcome it when it was forced upon them, as it was in the 1800s.

Fortified by trading and missionary zeal, the ever-expansionary forces of Europe sought in the 1800s to spread European influence into those few parts of the world where Europeans were still not known. China was among these places. The British, leaders in the expansionary endeavours, fought the so-called "Opium Wars" with China in 1840-42, and forced trading concessions out of China as a result. One of the most important of these concessions was the right of the British to create a number of trading "treaty ports" along the Chinese coast. It was thus that Shanghai, for instance, became an important town, providing a link between China and the outside world. Another very important link created by the British, and ceded on lease by the Chinese, was the port of Hong Kong. Other European countries, including Russia, Germany, Portugal, and France, followed the lead of the

Shang = up by, next to
Hai = sea

British, and inevitably the Chinese people became angry. They resented these intrusions into their traditionally esteemed way of life. So they rebelled. The Taiping Rebellions of the 1850s and the Boxer Rebellions at the turn of the century were the chief ones, but not the only ones. Eventually there was a revolution (in 1911) and Sun Yat Sen became the leader of a new republic. After 2000 years, the empire had been overthrown. But the change was only in the rulers; the people continued to live as their forefathers had done. Of course, after 2000 years the old forces of empire were not about to give up easily. The result was civil war, with Sun Yat Sen, and later Chiang Kai Shek, leading the republicans on one side and the war lords and imperialists on the other. Eventually the republicans won, and it looked as though peace would be restored. But it was not to be so.

The toppling of the empire had opened the door to other possibilities. One was communism. Another was invasion by an outside power. Mao Tse Tung had been influenced by the writings of Europeans, chiefly Marx and Lenin. (Marx was a German who lived and died in England; Lenin was a Russian.) He saw the strife in China after the destruction of the empire as an opportunity to spread the gospel of communism. The Japanese saw it instead as an opportunity to invade. They had deliberately moulded their economic-growth program upon the pattern of the West, and when Japan saw the Europeans with empires, Japan wanted an empire, too. China was close to Japan and in a weak state, and it offered a potentially secure source of lots of minerals. So Japan invaded. In 1931 they took over Manchuria, and in 1937 they started to invade the rest of China. With the communists and the Japanese both against him, Chiang Kai Shek did not have an easy time. By the time the Japanese had been beaten in World War II (and therefore also in China), the republican armies were in a sorry state. The communists soon took control, forcing the republicans under Chiang Kai Shek to retreat to the island of Taiwan. It was 1949. In not much more than a century since the Opium Wars, China had experienced a massive change in its rulers. After all, it had been the Opium Wars which first successfully challenged the authority of the emperors; from then on the Empire was in decline. It finally crashed in 1911, and after the dust of centuries had settled it was the communists who were in power. The stage was now set for

Louis Tsao

Family group in front of memorial to 67 rebels killed by government troops in 1927.

Walter Chin

A group of students at Kwongchow University in 1948.

a change in the Han way of life, which had existed for thousands of years.

TRADITIONAL CHINESE LIFE

The vast majority of Chinese were peasant farmers. There were very few merchants, because farming was almost all of the subsistence type, and the farmers rarely had any surpluses to trade away. There were some artisans, however, making pottery, clothing, tools, and ornaments. There were also the Imperial civil servants; occasionally a bright student from one of the villages would pass all his exams and be recruited into the civil service. He would then move away to a town. The towns were mainly administrative centres, the hands and arms of the emperor. Sometimes the hands were grasping, for there were tax collectors, too. And occasionally the arms were punitive, searching out wrong-doers for justice. But most of the people were farmers, plain folks.

They farmed tiny plots of land. As time had passed the number of Han people had grown, and so the land had become subdivided into minute holdings. Some of the population pressure had been relieved by taking some of the gentler hillsides fringing the valleys into terrace cultivation. Some had also been relieved by putting every available little bit of land to use (to the extent even that in some areas the people actually lived on boats in the rivers so that they would not take up valuable farmland). But mostly the pressure was unrelievable, and had been met merely by subdividing the land into smaller and smaller portions. The average size of farm in China in the 1940s was a meagre hectare, and while a few were larger (up to 2 ha) most were in fact smaller, as small as 0.2 ha. With farms this size, the Chinese were more like gardeners than farmers, but they survived.

In addition to being tiny, the farms were also scattered; a single farmer's holdings would not all be in one place. There might be a few square metres in one place, a few more square metres somewhere else, and maybe still more square metres in yet another place. As you can imagine, the scattering of the holdings made the problem of minuteness just that much more difficult. Yet the farmers survived.

The farmers had to deal with their small, broken holdings in an extraordinarily intensive manner in order to get enough food from them to feed a family. Work was

Sue Mason

Artisans at work in Peking.

Hei = black
Pai = white
Hwang = yellow
Hung = red

Pei (or *Peh*) = north
Nan = south
Tung = east
Si = west

long and arduous, often taking up the entire daylight hours. Cultivation was practised very carefully; plants were intercropped, and if possible 2 harvests were taken each year. Everything was done by hand, and nothing was wasted. Soil fertility was kept as high as possible both by repeated river flooding (bringing silt) and by the use of every type of fertilizer available. The fantastically intensive use of the land in this manner yielded enough for the farmers and their families to live on — but only just. Famine was a constant danger and an occasional visitor.

Animals are relatively wasteful users of land compared with crops. For example, 1 ha of rice can provide as much as 10 to 15 times as much food as the same hectare would if used for grazing cattle. So the only cattle the Chinese kept were those necessary for pulling ploughs and wagons, namely oxen and water buffalo. Needless to say, there were not many about. Pigs, ducks, geese, and chickens were much more common, the reasons being that they can feed off left-overs and other scraps that people cannot eat and they do not take up much room. Pigs sometimes even lived in the same huts as the people. Both pigs and fowl are fairly efficient at turning scraps and other waste into nutritious food, so they were ideal for the Chinese. The Chinese also obtained quite a lot of valuable protein from fish, which could be kept in the rice fields when they were under flood.

The farmers were obviously very ingenious and very efficient at making a survival living from the soil. But that was all it was, a survival living — no extras. Notice, too, it was *the* soil, not *their* soil. The great majority of the farmers did not own the land they worked; the land was owned by landlords, sometimes living in the village, sometimes living far away in the town. Sometimes the land was in pawn to the moneylender; sometimes it was under confiscation. Only rarely was the land owned freely by the farmer who worked it.

Such was the traditional pattern of Chinese life on the eve of the communist take-over of the government. There was not much in the way of industry and urban life. Nevertheless, some cities existed — chiefly the administrative centres of the old empire and the old treaty ports of the European and Japanese ingressors. What industry there was existed almost entirely in the treaty ports and Manchuria, for it had largely been set up by the Europeans and Japanese. Furthermore, it was almost entirely connected with either transportation (e.g., railway work-

Hope McBride

Baskets at the Shanghai Basket Fair, 1936.

Kow (or Kou) = mouth of a river
Shan = mountains
Ling = peak
Hu = lake
Kiang = river
Ho = river
Chwan = river
Yun = cloud

Tien = heaven
Ti = earth
Lung = dragon

Hsien = country
Chow = district

People's Republic of China

A record from 1812 saying that despite the very bad harvest the rent should be collected from the peasants as usual.

Sue Mason

A commune worker.

shops) or with the processing for European markets of goods produced inside China (e.g., silk and cotton textiles). Manchurian heavy industry was an exception. Shanghai was by far the most important industrial centre, and also by far the largest city. No-one knew exactly how many people lived in the cities, but several were over a million in size. Shanghai, Peking, Tientsin (look at your atlas again), Shenyang, Wuhan, Chungking, Canton and Nanking were all probably over a million each, big by world standards, but unimportant by the dominant farming standards of China. Perhaps only about 10% of the total population lived in the cities.

Then, on October 1, 1949, the communists proclaimed the establishment of a new state — the People's Republic of China, or as they call it, Chung-Hua Jen-Min Kung-Ho Kuo.

THE COMMUNISTS AND FARMING

Four thousand years of traditional farming disappeared almost overnight. The extent and rapidity of the change were incredible. Starting in 1950 the landlords were all totally dispossessed of their land. Every part of it was confiscated by the state. The first phase of the state's action was not complete, however, until all the land was shared out among the millions of landless peasants. The re-allocation of the land to the peasants was not completed until 1952, by which time about 500 000 ha had been redistributed to over 300 000 000 farmers. The farmers were ecstatic. They owned their land — no more landlords, no more moneylenders.

King = capital city
Kwan = supervisor

Chung = central
Kuo = nation, country
Chungkuo = central country/China

Chung-Hua Jen-Min Kung-Ho Kuo = Central Flowery People's Republican Country

After having broken up some of the large landholdings into smaller pieces, the communist organizers decided that it would be a good idea to put some of the pieces back together again. How can they put back into a *square shape* the 5 different holdings belonging to the Green Dragon co-operative?

However, the state had not finished its reorganization. Phase 2 had still to come. Starting in 1953 the millions of peasants were grouped into co-operatives, whether they wanted to or not. But the idea made sense to most. After all, why not share the few oxen and water buffalo around? Why not share the precious fertilizer? Why not share the wagons? Why not help one another? The co-operatives consisted at the lowest level of about 10 to 15 families, who shared the day-to-day work among themselves. These family co-operatives were then grouped (10 to 15 at a time) into larger co-operatives, called producers' co-operatives. By 1956 nearly a million producers' co-operatives had been organized, and 90% of all the land under cultivation was accounted for. All the farmers were allowed to retain very small plots for their own use, regardless of the decisions of the co-operative. For the rest of the land the farmers worked as a single team and took out in payment according to the work they had contributed. Notice that no-one now *owned* the oxen, the water buffalo, the wagon, the implements; these had all been confiscated for *public* ownership. The only thing left that the individual farmers could contribute was indeed their own labour. For this, they were paid.

The farmers did not foresee phase 3. In 1958, in a tremendous rush, the state organized the producers' co-operatives into communes. Before the end of the year, some 25 000 communes had been set up, each consisting, on average, of about 7500 families, composed of the members of 50 or so producers' co-operatives. The land under the influence of a commune extended for about 500 ha, within which there would be several villages. The commune organization took over many of the old village functions — welfare, protection, education, and management. Production was organized on the basis of "brigades", each consisting of several "teams". Thus there would be a rice-producing brigade, with many teams; there would be an irrigation brigade, with several teams; and there would be other brigades for all the other activities within a commune (such as power generation, garbage disposal, construction, education, health care, and so on).

The great advantage of the brigade system is that it permits each person to specialize in doing the thing he does best — whether it be nursing or construction, electrical work or tending animals. This specialization is called *division of labour*. It generally causes an increase in production, because the people are not constantly moving from one thing to another during the day, thus wasting

Trace out the following figure. Now, can you divide it into 4 areas, all with identical shapes and sizes. That's the problem the Central Committee faced when it came to divide Heung-ming Kwan's estate.

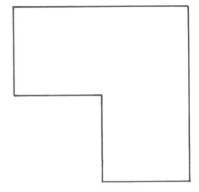

In one of the communes in central China there were 16 brigades who tended to get in one another's way. The commune decided to keep them all apart by scattering their offices as shown below and then dividing the Commune by just 5 straight lines. Can you trace off the brigade offices and divide them, each from the others, by using just 5 straight lines?

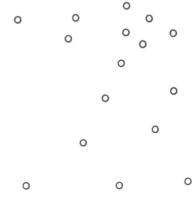

A society in change.
All photos: Sue Mason

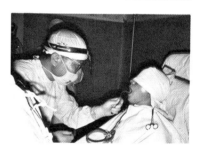

time. In consequence, people tend to be better off. However, the commune system was not designed to allow people to remain in just one activity for ever, because the communists did not see that as a way to complete social equality. Equality would be gained only if everyone did everything, though at different times, of course. Thus a person may serve as a rice grower for a season, then become a construction worker, then act in the following year as an irrigation worker, and so on. If a person wants to develop a particular skill to a higher level, necessitating study at college or university, then part of the price of the right to study is compulsory work in the fields and on the construction sites. It is only in this way, say the communists, that the future managers and "white-collar" workers can understand what it is like to be a worker in the field.

By the end of 1958 the commune system was in full operation. In less than 10 years, then, the entire way of life of the Chinese farmers had been swept aside and replaced by something unique in the history of the world. Why did such a huge change succeed so quickly? No-one can really know the answer to that, of course, but we can probably accept several suggestions. First, the peasants were totally dissatisfied with the traditional pattern of landlords, moneylenders, debt, poverty, hunger, the risk of famine, and the desperate need for constant hard work. They were therefore willing to listen to the communist suggestions for ways to improve matters. Second, the Communist Party workers (only about 100 000 strong in a total population, during the 1950s, of about 550 000 000) were highly motivated to succeed. They had fought for control of China for years, and eventually gained it; they were not now going to sit back and relax. Third, the communist government was committed to a marked improvement in the standard of living of the people of China. It *had* to try to do something better than the old empire had done, if only to consolidate its victory over Chiang Kai Shek (who would probably have invaded mainland China if only he thought the mass of the people would have turned against the communists). So willingness on the part of the farmers and motivation and determination on the part of the communists combined to produce almost unbelievable change.

There have been few changes since 1958. Indeed the structure of the commune system could hardly be expanded, for 99% of the rural population was already

Sue Mason

Getting to work in Peking.

The medical team were in office A, the rice-growing team in B, the iron research group in C, and the irrigation managers in D. They each had to get to work from their separate apartments in blocks A, B, C, and D respectively. Since they all had to be at work at exactly the same time, and they didn't want to get in each other's way, they devised a system of routes they could take that did not cross one another. Trace out the diagram below; can you join A to A, B to B, C to C, and D to D, using 4 lines that do not cross one another?

Apartments

Offices

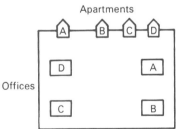

Sue Mason

Specialized kindergarten care on a commune.

within it. But what did happen was that the authorities decided that the communes were too large. So it more or less cut them into thirds, thereby tripling their number to about 75 000. The only other significant event of note was the so-called Cultural Revolution, which began in 1966 and is not really over yet. The farmers, as well as many other people throughout Chinese society, did not all take kindly to every idea proposed by the communists. Mostly, of course, they kept their dissent to themselves, because the communists were the ones who had fought in the great civil war against the republicans; the communists were soldiers, often tough and not concerned to argue fine points. The communists were also the government, and that still meant a lot to the Chinese people; 2000 years of obeisance to emperors had not prepared many of the Chinese for arrogant independence of mind. So the farmers, along with others, did what they were told. However, they often did it half-heartedly. They sometimes muttered among themselves; they almost complained. Realizing what was happening, in 1966 Mao Tse Tung instituted the Cultural Revolution. Its purpose was to remind all half-hearted people that communism was really here to stay, and that all traditional ideas were dead. The traditional "culture" was finished; there had been a cultural revolution. Students who favoured the new against the old were recruited as "Red Guards" to go around the country and remind everyone that communism was permanent. They were instructed to "denounce" people who clung to the old ways and old ideas. These denounced people were called "revisionists", because they wanted to revise the revolution that had occurred. Some of them were actually killed; many were fired from their jobs; a few escaped. The farmers survived.

No-one really knows exactly what happened to all the revisionists. The official Chinese News Agency does not release information unless it expects that the rest of the world will somehow gain an even higher regard for China. There is no censorship as such, in that news items are not deleted from a reporter's copy. It's just that the issue of news is carefully controlled, and not much is allowed to be reported in the first place; so there is no need to delete passages.

THE COMMUNISTS AND INDUSTRY

The first Five Year Plan (1953-57) stated that it was one of the long-term aims of the new government to "transform China from an agricultural into an industrial nation." At first the plan went smoothly, mainly because the Chinese got enormous help from the Russians. The Russians had developed plenty of expertise in the setting up of a communist industrial state, and they actually wanted very much to help the latest recruit to communism. The U.S.S.R. was especially helpful in setting up heavy industry, such as mining and iron and steel. This happy state of affairs persisted for quite a while, but eventually the Chinese and Russians began to disagree about priorities — about what should be done first, and what should be done next, and so on. The Russians therefore withdrew.

In 1959-60 the second Five Year Plan, which had started in 1958, collapsed. The Chinese claimed publicly that there had been a "Great Leap Forward", but in practice they went on quietly trying to pull industrial growth back out of the mess it had fallen into when the Russians withdrew. Throughout the 1960s and into the 1970s the Chinese have not published any data on industrial progress, so we cannot know exactly what has been happening. Planning appeared to continue on a yearly basis, despite the announcement that a third Five Year Plan was to start in 1966 (the year the Cultural Revolution started). In 1971 a fourth Five Year Plan was announced, but no-one knows much about it.

One thing is certain. Industry is not progressing in quite the same manner as it did in Russia, nor even in the West. Local industry is getting priority treatment, and the communes are being strongly encouraged to become as self-sufficient as possible. If this means that a particular commune needs, say, a steam forge for its agricultural implement making, then the commune is encouraged to make one. If you need something, make it — that's the philosophy. One obvious reason for this, of course, is that the Chinese still do not have as much money to buy things as they would like. So — make the things you need. If you can't, then go without. If the things are too important to go without, and they cannot be made in the communes, then they can be bought from the towns. If really necessary indeed, then they can even be bought from another country; but that is strictly the last resort. The major result of this policy of communal self-sufficiency as far as possible in industrial goods has been to scatter industry throughout the country. There are no parts of China now without at least *some* industry, even if it is only small-scale local industry.

The dispersal of industry is also a major policy aim of the communists in the field of heavier large-scale industry. Prior to the communist revolution most factory industry (such as it was) existed in the eastern coastlands and in Manchuria. The communists regarded this as a very uneven sort of distribution pattern, and they wanted to have

Do it yourself on a scale involving millions of people! Beaver Lumber and Home Hardware would make a fortune!

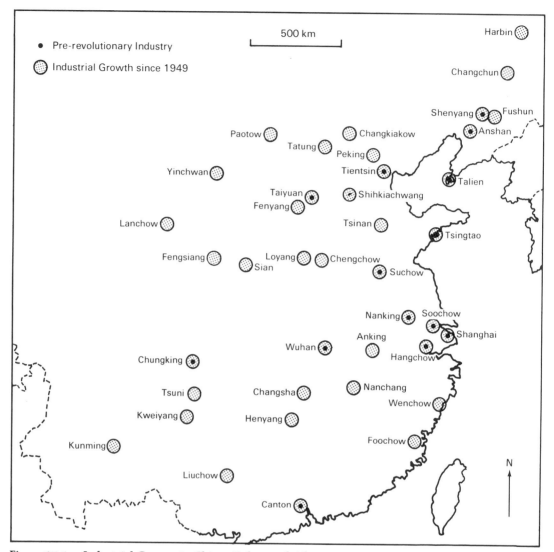

Figure 17-1.   Industrial Centres in China: Before and After 1949

more factories in the interior — partly for reasons of equality of work experience, but also partly for strategic reasons (it was regarded as safer in the interior). Figure 17-1 shows the pre-1949 distribution of industry, and compares it with the communist distribution.

THE COMMUNISTS AND COLONIZATION

The distribution of population in traditional China is shown in Figure 17-2. For the most part the population

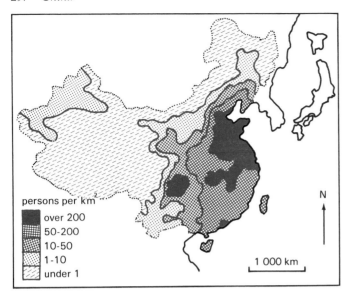

persons per km²

- ■ over 200
- ▨ 50-200
- ▧ 10-50
- ▨ 1-10
- ▨ under 1

N

1 000 km

Figure 17-2.   Generalized Population Distribution of China

was concentrated in the river valleys and plains of the eastern coastal areas. The communists regarded the relatively empty remainder of the country as offering potential wealth, plus protection from possibly aggressive neighbours. The empty parts just needed more people. Accordingly, they stepped up the already existing drift of people into Manchuria. Indeed, the drift became a surge. The only point is that the word "Manchuria" does not exist in Chinese administration; to be proper you must really call it the North-Eastern Region of Heilungkiang, Kirin (which is also the name of a Japanese beer), and Liaoning. Manchuria is probably easier.

Sinkiang-Uighur (look at your atlas!) was another focus of Chinese expansionary interest, lying as it does right under the heart of the southern U.S.S.R. It is the place that the Chinese now use — in its desert parts — for testing nuclear bombs. How do you think the Russians feel?

The Chinese also consolidated their western borders by moving into the mountains as well as into the deserts. Tibet, which had been part of China before the empire became weak, and which had broken away after the empire fell, was brought back into the nation in 1959, when Chinese troops invaded. Since then as many as 500 000 Chinese have been placed in Tibet as colonists, and the commune system has been extended to include them. Tibet is now once again effectively part of China.

THE COMMUNISTS AND THE POPULATION

The first (and so far only) complete census of the population took place in 1953, and revealed that there were then 583 000 000 Chinese. The census was a sort of communist stock-taking. Before that there were no exact figures, merely rough estimates by various bodies. The estimates give an 1800 population of 350 000 000, a 1940 population of 400 000 000, and a 1950 population of 500 000 000. Keeping track of the population since 1953 has been an unenviable task, performed by the U.N. The U.N. estimates that the population had reached 670 million by 1963, just 10 years after the census, and had risen further to about 800 million by 1973. The Chinese do not deny these figures, but then they probably do not know exactly either.

In the early years of the People's Republic, the government actively boasted of the large numbers of people, regarding a large population as an economic and a strategic asset. In time the daunting problems of development were more fully understood, and the government began to take the line that a population that was growing *more slowly* would be a good idea. Thus the government began a massive birth control campaign, which worked — not surprisingly, because the state is organized in such a way that what it decides is *made* to work. Birth rates are now down everywhere in China (as are death rates), and China's population growth is at last beginning to slow down. The problem is not solved yet, of course, and the state still needs to keep up its birth control pressure. For example, late marriages are encouraged; food tokens are given for no more than 2 children in a family; birth control advice is freely available: posters are everywhere, and "barefoot doctors" preach the benefits (by the way, each production "brigade" has its own barefoot doctor and clinic). Birth control is now part of the cultural revolution.

Another way in which the government is attempting to control population is to try to move it out of the cities. In all other countries of the world, people are drifting to the cities (see Chapter 20) and causing all sorts of problems. In China the government does not let this happen. Indeed, it is even trying to get the movement to occur the other way, and with some success. For example, it is now compulsory for a person who wants to gain promotion or to go to university to have to spend a year or so out in a commune. Students at school are also sent in large gangs to help work in the communes — 6 months at school, 6 months in a commune. (No holidays, notice.) Part of the reason for these actions is that housing is extremely scarce in the towns; so are jobs. For the time being, a largely rural population is much better — there are jobs; there are houses (or space to build); there is a need for food; there are fewer opportunities to develop revisionist tendencies.

Revisionist? look back!

The organization of a state to produce this sort of effect in a short period of time, as the Chinese have done, is clearly made easier when everyone speaks the same language. In traditional China, people spoke several different languages; the written characters were the same, but the spoken words were different. The most common language, spoken by 75% of the population, was Mandarin. It was spoken everywhere in the west, centre, and north. Only in the south were there different spoken languages: Cantonese, spoken by about 10% of the population, and Min, Wu, Xiang, Kejia ("guest people"), and Gan. In keeping with the communist policy of uniformity of experience and information, these local languages are gradually being replaced by Mandarin. And in keeping with the idea of minority rights, these local languages are also being sponsored for preservation as part of a minority culture. In other words, people are being forced to learn Mandarin, yet are also being given the choice of maintaining their own minority language.

THE COMMUNISTS AND THE STANDARD OF LIVING

It was the never-ending grinding poverty and total hopelessness of the farmer's life that largely favoured the victory of the communists over the republicans in 1949. The communists have had a debt to repay, and they have repaid it. There is now no doubt that life for the peasants is immeasurably better than it was under the traditional regimes. For many years the Chinese released no information about production, and people from the outside could not easily gain access to China itself. This changed in 1970, when Canada opened up diplomatic channels with the Peking government, followed shortly after by the U.S.A. and some European countries. Visitors from the outside were now welcomed; tours were arranged, and certain communes were even set aside as tour highlights. Precise production data were still not available, but at least it became possible for visitors to form opinions and gain impressions. On the basis of various observable facts, economic experts in the West estimated that China's economic growth since coming under communist control had been of the order of 10% or so a year, making it about as fast a growth rate as Japan's.

Some of the impressions brought out by visitors (1970-73) include the following (bear in mind that you must compare these impressions with pre-1949 Chinese hunger and poverty and not with post-1975 Canadian wealth):

— rice 20¢/kg, cabbage 6¢/kg, pork 70¢/kg;
— suit (standard blue cotton jacket and pants) $7 and a year's supply of cotton coupons;
— *Phoenix* bicycles $75 each; *East is Red* transistor radios $15 each;
— *Shanghai* wristwatches $35 each; sewing machine $90; shoes $8;
— monthly rent on a tiny house $2 for a family (single persons do not get houses);
— monthly rent for a bunk in a factory dormitory 50¢;
— average monthly factory wages $22 to $28, depending upon skill and experience;
— hours, 8 a day, with half-an-hour for lunch, 6 days a week;
— annual vacation, none;
— workers are assigned to factories, and may be re-assigned later; there is no right to leave a particular job for another; there are no strikes;
— farmers share the income of the commune; average personal income in a commune is between $5 and $6 per month;
— Mao Tse Tung badges 10¢ each;
— newspapers posted on walls for people to read;
— cars, washing machines, refrigerators practically unknown to the peasants and factory workers;
— Chinese airlines fly some Ilyushin jets supplied by Russia before the Russians withdrew in 1960; the jets are probably not very safe; maintenance is minimal;
— Chinese people smoke a lot of cigarettes (so does Mao); cancer is common;
— they also eat chopped dog for medicinal purposes; its price is 0.83 Yuan per catty (about 75¢/kg);
— most food is rationed; people carry ration books every time they shop;

— fish is rationed to 2 catties per person per 10 days (a catty is 0.5 kg);
— the communists claim that prices have not changed since 1949;
— the standard meal for a peasant or a factory worker is a bowl of rice with a few pieces of fish or meat, plus vegetables, on the top;
— meat is sold alive in the markets, and then slaughtered at home; there is very little refrigeration;
— revolutionary music plays in many places, even in the streets and parks;
— candy costs $2/kg; there is a lot of it;
— girls go to school now; they didn't before 1949;
— men and women earn equal pay for equal work;
— happiness is (in order) a camera, a bicycle, a sewing machine, a radio; the cheapest cars cost $6000;
— nursery-school children package small industrial components as part of their play;
— schoolchildren learn that laziness is counter-revolutionary.

QUESTIONS AND EXERCISES

1. Read *A Memoir of China in Revolution* by Chester Ronning, Canada's former ambassador to China. The book was published in 1974 and contains some of the post-1970 open-up impressions as well as some of the traditional views. It's good for gaining the idea of a society that has changed.
2. Research the Hwang Ho project. It is a massive river-basin project, akin to the T.V.A. scheme in the U.S.A., and you should be able to find out something about it fairly easily. Don't be discouraged: the Chinese aren't.
3. What are the general lines of development that you think China still needs to follow if progress is to be maintained? Or is the present organization of society the final product?
4. Find out what you can about the Chinese fishing industry.
5. Compare the climates of China north and south of the Wei Ho Valley (*Ho* means "river"). What differences do you think these climates will make in the types of crops grown in different parts of China?
6. Loess — what is it?
7. How is Hong Kong different from the rest of China?

18. JAPAN'S SOCIAL GEOGRAPHY

CHANGE

The Emperor Meiji, who came to power in Japan in 1868, was the man who set Japan on the long road to modernity and prosperity. In the century or more that has elapsed since his rule began, great changes have come to his native land. Many of the changes he would applaud, because they show that Japan of the 20th century has fulfilled his hopes and dreams.

"The Japanese are already the third greatest economic power in the world and could easily become the global leader by the year 2000."

"Japanese cars have over 50% of the imported car market [in the U.S.A.]. This market position once belonged to Volkswagen."

"Japanese . . . an amalgam of discipline, desire, hard work, skill and innovation."

But the Emperor Meiji would perhaps be equally saddened at other facets of Japanese life.

". . . age old Asian concept of family life is crumbling . . . a record number of child runaways . . . pollution poi-

Japan Information Centre
Consulate General of Japan, Toronto
This view of a fishing village in a rain shower was sketched over 100 years ago.

To'dai (TOkyo, DAIgaku) is Tokyo University, Asia's most important university. Started in 1877 by Emperor Meiji, it originally turned out government officials. Because its graduates were allowed to go into the civil service without taking an examination, Todai assumed an air of elitism. To get into Todai, one must pass a hard entrance examination for which parents often start preparing their children when they are only toddlers. Current enrolment is about 18 000 students, and about 3000 bachelor degrees are awarded annually. Graduates have a rosy future, for, often in complete disregard of the academic record there, a company will hire a Todai graduate on the basis of his simply having been there. Just getting through the entrance examination means that the student has ability!

soning blamed on the government's high-growth economic policies . . . traffic congestion . . . mass rural exodus
to the cities."

Before Meiji came to power, Japan had remained isolated
for centuries. It pursued a conscious policy of isolationism
that was finally ended by Commodore Perry of the United
States Navy. Literally at gunpoint, the Japanese had to
negotiate a trade treaty with the U.S.A. The old order, in
which a peasant who failed to avert his eyes at the approach of a samurai would be liable for instant execution,
was about to change. The emperor saw that it would be in
Japan's best interest to admit at least some Western influences in order to remain strong and independent. Most
of all, Meiji wanted to help his people. They were poor,
oppressed by warlords, and their numbers were growing
every year. Time was not on Japan's side.

In 1900, Japan's population numbered 43 000 000. By
1970, that number had increased to 104 000 000. On an
island group of only 372 000 km², one-thirtieth the size
of Canada, the average population density is 280 people
per square kilometre. In the cities, where some 76 000 000
Japanese live, the density is even higher.

So many people! You can be sure that any increase in

Japan Information Centre
Consulate General of Japan, Toronto

Himeji Castle, built in 1609, is rep-
resentative of the splendour of
feudal Japan.

Japan's population is something that is keenly felt. At least 80% of the archipelago is mountainous and unsuitable for settlement or cultivation. Farms are very small and average slightly more than 1 ha in size. In the cities, people are crowded into relatively small areas. While this has resulted in the enviable national characteristic of serenity, acquired over the centuries, the Japanese have been trying to solve their population problem for 30 years — ever since the end of the Second World War, in fact.

Year	Population
1900	43.0 million
1910	49.2
1920	55.9
1930	63.9
1940	72.5
1950	83.2
1960	93.4
1970	103.7
1975	109.9
1980	116.0
1985	120.8 } estimated figures
1990	124.7
1995	128.3

Figure 18-1.   The Growth of Population in Japan

With such a keen awareness of the necessity for limiting the size of the population, it is vital that Japan find out just how many people it has. Like most modern, industrialized societies, Japan holds a census every 5 years. A full one is held in the first year of every decade, while a smaller one is held every fifth year in each decade. For example, in 1960 a full census was held while the smaller census for that 10-year period (ending 1970) was held in 1965.

The most important problem that faced the Japanese people after the Second World War was the necessity for birth control. Death control had allowed people to live longer lives and babies were still being added at — then — an increasing rate. The baby "boom" that most countries involved in the Second World War experienced was felt in Japan.

As you can see from Figure 18-2, a great surge in births began after the Second World War, at a time when the death rate continued to decline. You can also see that in the 1920s and the 1930s, the birth rate was declining, yet the population continued to grow. Even though the birth rate had dropped a little, the death rate dropped

Japan Information Centre
Consulate General of Japan, Toronto

Sakuma Dam. The generation of hydro-electric power is highly developed in Japan, taking advantage of the abundant rainfall and many swift-flowing rivers. The Sakuma Dam is one of the largest in the country.

Year	birth rate per thousand population	death rate per thousand population
1921	35.1	22.7
1931	32.1	19.0
1941	31.8	16.0
1947	34.3	14.6
1951	25.3	9.9
1955	19.4	7.8
1960	17.2	7.6
1965	18.6	7.1
1970	18.8	6.9

Figure 18-2.   Japan's Vital Statistics: Birth and Death Rates for Selected Years

much more! After the Second World War the rate of natural increase was even higher than during the 1930s.

It has long been the policy of the Japanese government to encourage birth control. There are other reasons why the birth rate is falling, however. Many large organizations offer free birth control information to their employees in order that as few women as possible will leave to have families. Prosperity has encouraged many potential parents to have small families or no families at all. Such people find themselves able to afford many of the new consumer goods, such as cars and TVs, that Japan produces in such quantity and quality. Programs of social welfare no longer stress a large family as an insurance against old age and the time when one can no longer work. Now there are pensions. This is very important,

Age group	Year		
	1950	1960	1970
0–4	13.5	8.4	8.5
5–9	11.4	9.8	7.9
10–14	10.5	11.8	7.5
15–19	10.3	10.0	8.7
20–24	9.3	8.9	10.3
25–29	7.5	8.8	8.7
30–34	6.3	8.0	8.0
35–39	6.1	6.4	7.9
40–44	5.4	5.4	7.0
45–49	4.8	5.1	5.7
50–54	4.1	4.5	4.6
55–59	3.4	3.9	4.2
60–64	2.8	3.1	3.6
65 and over	4.6	5.9	7.4

Figure 18-3.   The Changing Age Structure of Japan's Population. For each year, the percentage of the total population is given by age group.

Japan Information Centre
Consulate General of Japan, Toronto

Single-family detached homes are rare in Japan. Why?

Japan Information Centre
Consulate General of Japan, Toronto

The 1974 Spring Labour Offensive. Workers parade through the streets as part of their annual fight for higher wages. Nearly all unions make their wage demands at the same time — in the spring.

because life expectancy in Japan is now over 70 for both men and women, when before the war (1939-1945) it was less than 50 years.

How do all these people manage to make a living in a country that is nearly 60 times more crowded than Canada? And which has negligible natural resources, so that the Japanese economy depends almost completely on imported raw materials from places like Australia and Canada, and export markets all over the world?

For most of the last 20 years or so, the Japanese have had to cope with between 1 000 000 and 1 500 000 people coming onto the labour market each year. Even so, the Japanese have been able to provide almost full employment for their people. (Perhaps you can see why the 1960s were especially difficult for young people trying for their first job in Japan. Remember, the school-leaving age in Japan is 15.)

Many people believe that the Japanese are in the habit of exporting nearly everything that they make. This is not true. Japan's people have been experiencing a tremendous surge in buying power, even allowing for inflation, which reached 24% in 1973. In fact, Japan's exports were at their highest level ever in 1970, when only slightly more than 10% of Japan's goods and services produced in that year were exported. Most of the tremendous increase in Japan's economy (more than 2000% between 1950 and 1970) has been absorbed by its own people.

Today, more than 75% of Japan's population is of working age. Pessimists are saying that Japan is likely to grow less quickly economically in the near future. The pessimists have been examining the age structure of Japan's population and have found that the population is gradually becoming older.

Put yourself in the position of Shigeo Nagano, President of Fuji Steel, or sit on the tatami of Fumihiko Kono, head of Mitsubishi heavy industry. They are worried men, because the percentage of the population that is in school is declining. If the economy is to grow, where are the workers to come from? The answer can only lie in the increased automation of industry and the phasing out of unnecessary jobs. In other words, productivity has to rise. Yet if it rises, the people might become more affluent and consequently have fewer children. Thus, the problem of fewer workers replacing the ones who die or retire might become worse. A solution might be for Japan to "import" workers. Yet the country is already very crowded.

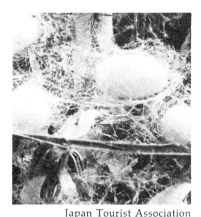

Japan Tourist Association

The silkworm cocoons above are unravelled and woven into the richly patterned traditional kimonos worn by the women below. Silk kimonos cost the equivalent of hundreds of dollars.

Japan Information Centre
Consulate General of Japan, Toronto

The age of retirement for many Japanese workers has been increased from 55 to 60 years of age.

Nippon steel is the world's largest steel company. It produced over 40 000 000 t of steel in 1973, more than the output of some entire countries. Among the many workers are 2 men, one old, one young. The old man recalls having to wash his older comrades' clothes when he was an apprentice at 14. The deference shown to one's elders is still continued today, he says. Younger members generally bring tea for their elders on their lunch break. The young man nods in agreement. He earns $350 a month, which compares well with the $8 a month the older man earned when he started first. Both men live in subsidized apartments, owned and rented by the company. It costs about $15 a month to rent one. Both men feel that working for Nippon Steel is almost a family affair and are prepared to stay for the rest of their lives. They are confident that Japan can continue to produce steel more cheaply than almost any other country, although the 1973 oil crisis shook their confidence a little. They feel quite assured about the future when they contemplate the tight links between industry, government, and the banks. Pollution worries them, but they are confident that their country will solve the problem in the near future.

In 1972 Japan's labour ministry reported that there was an estimated shortage of 1 359 900 workers and that even with automation replacing manpower, a shortage was likely to persist. Such a shortage has resulted, inevitably, in higher wages being paid. Wages have gone up at least 200 or 300% since 1960 and are likely to continue rising. While inflation has eaten into the value of money, in real terms the Japanese worker is much better off than he was in 1960.

The labour shortage in Japan has resulted in the trade unions demanding higher pay. Usually this takes place in spring in a process called *shunto*. It begins on March 31, timed to coincide with the end of the financial year. Many unions will strike at the same time, often paralyzing the country in several respects for a while. This has had the effect of forcing many small firms out of business. They simply cannot afford to pay current wages and salaries, cannot attract workers, and so go broke. Industrial mergers are resulting in the large companies becoming larger all the time. Small companies, like food stores, are so short of help that they are cutting out home delivery. Many construction companies go bankrupt every year.

In addition to a shortage of workers, there is a growing trend, especially among the young, to be dissatisfied with tedious, assembly-line occupations. In 1972, 34% of all workers in the 20-24 age group quit their jobs. They were dismayed at what they considered to be socially worthless jobs, rather than the rate of pay they received. This is a great change from 1945, when the 7 000 000 returning soldiers went job hunting and expected to work for one company for life. It was life-time security and, in return, the worker might be expected to give his entire loyalty to his employer. Until quite recently, to have had experience with several different firms was indicative of disloyalty. In the Western world, such a work record is regarded as a positive asset.

The growth of Japanese industry has been phenomenal since the Second World War. It has been largely based on a huge, increasingly affluent home population. Only some 10% of Japan's goods and services are exported. The demand for labour has been such that more than half the women of Japan hold full-time jobs. That resource has been fully utilized. The attitude of the Japanese to work has traditionally been one of great seriousness. In 1972, more than 80% of all workers did not take the holidays to which they were entitled, and a Tokyo survey showed that only 2 people in every 5 showed any desire for a 5-day work week. Assembly lines are commonly run for 14 hours a day, 6 days a week.

Accompanying the trend to increasing industrialization has been the migration of many people from the countryside to the city. They moved in the expectation of a higher standard of living, which most fulfilled by taking a job in a factory or working in service industry. Many Japanese farm owners took a job in the city while their wives

Dengo Koshuuko works for a Japanese car firm well-known internationally for its products. He earns about $400 a month and receives an annual bonus of up to 6 months additional salary, like many other Japanese. He is in charge of an assembly line, likes to arrive early for work in the morning so that he can be on top of the job by the time the other workers arrive, and rarely leaves when the official working day is over. Lunch is eaten in the company cafeteria, where food is very cheap. Dengo has advanced to his present position solely by staying with the one company and being promoted on the basis of seniority.

Most Japanese companies have some form of calisthenics for workers during the day.

The *zaibatsu* are the largest combines in Japanese industry. The big three are *Mitsui*, *Mitsubishi*, and *Sumitomo*. They produce a bewilderingly large variety of goods and services, and scour the world for raw materials. Mitsui alone sells goods and services worth nearly $10 billion annually.

The Ministry of International Trade and Industry fines its people if they come to work on their scheduled holidays. The Mitsubishi shipyard in Nagasaki closed its gates in 1974 for Saturday work. Even so, the first time it happened, 400 workers turned up and chanted slogans outside the gate, demanding to work. One executive said that he felt the extra day of imposed idleness would make him soft!

New Japan now dawning
With industry helping her grow,
Happiness in hard efforts
Shared by all,
Our enthusiasm fills
Matsushita Electric!
The company song of Matsushita Electric, makers of Panasonic products.

The Tokyo Maru, a 150 000 tonne oil tanker, is being launched.

A Japanese cultivator being demonstrated in Morocco.

Steel is one of the principal pillars supporting Japan's economy and its industries. It has become the world's third largest producer of steel.

Assembly line worker in a car factory.

Cultured pearls are a specialty of Japan.

worked full-time on the farm. This was not so difficult as it sounds, as most Japanese farms are very limited in extent and resemble large, very carefully tended gardens.

Year	Primary	Secondary	Tertiary
1950	48.3	21.9	29.7
1955	41.0	23.5	35.5
1960	32.6	29.2	38.2
1965	24.6	32.2	43.0
1970	16.8	34.0	49.2

The above figures are percentages of the working population engaged in the 3 main branches of economic activity.
A *primary worker* in Japan is usually a farmer, or a fisherman.
A *secondary worker* in Japan is usually a factory employee.
A *tertiary worker* is someone who provides a service. In Japan such a person might be a salesman, a waiter, or a sales clerk. Generally, the tertiary worker lives in a city.

Year	Urban population (%)	Rural population (%)
1950	37.5	62.5
1955	56.3	43.7
1960	63.5	36.5
1965	68.1	31.9
1970	74.8	25.2

Figure 18-4.   An Aspect of the Changing Demography of Japan

As you can see from Figure 18-4, nearly one person in every 4 is a rural-dweller. This is still quite a high proportion (25%) when compared with Canada (11%) and the U.S.A. (12%). In other words, urbanization in Japan still has some way to go before it reaches the proportions of other parts of the industrialized world. This has led some Japanese to say, "It is impossible for any more people to pack into the cities. It will be impossible to house them, feed them, and provide them with jobs." The crowding in urban centres like Tokyo is apparent when you think of the people employed to push commuters into the subways at rush-hour. Yet Tokyo (population 11 000 000 in 1970) is still growing and linking up with other cities. By 1980 it is believed that Tokyo will be only part of a great megalopolis that will house at least 75 000 000 people. That figure is more than 3 times the population of Canada. Even though the island of Hokkaido is now being opened up for development, the fact remains that Japan has very little room for further development. Even though a road and rail tunnel now connects

Japan Information Centre
Consulate General of Japan, Toronto

Volcanic islands emerge from the sea, sometimes disappearing at once, sometimes growing in size. Japan is situated on the Pacific Ring of Fire, an earthquake belt stretching around the Pacific. Sights such as this are not particularly rare.

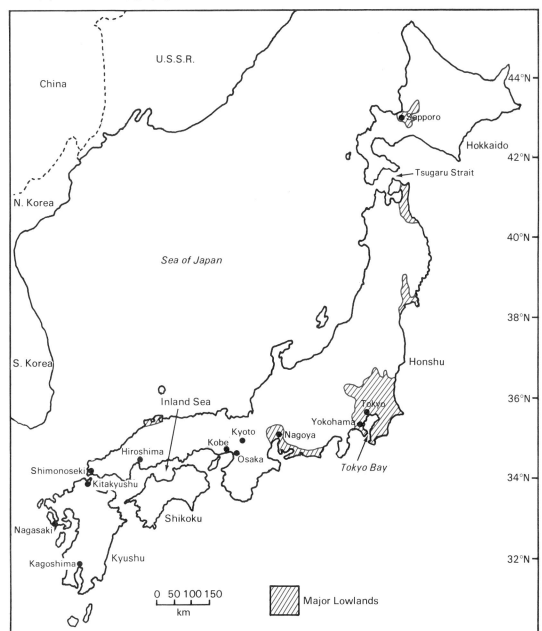

Figure 18-5. Japan

Hokkaido with Honshu, it is hard to see how this will
alleviate the problem of urban crowding in the south.

It should be remembered that Japan has not just suffered from the disadvantages of too many people and a
high rate of pollution. (Did you know that more than half
the people of Tokyo are not served with main sewage,

and that night soil is collected early in the morning by tanker trucks, taken to the harbour, where it is loaded onto the "honey fleet" for dumping on islands farther out to sea?) There has been the advantage of economic gains — of full employment and rapidly rising standards of living, due to having a large home market that allows industry to practise economies of scale.

Year	Telephones in use	Television production	Radio production	Cars in use
1953	2.6	0.01	1.4	0.1
1964	12.3	5.3	25.6	1.7
1965	14.0	4.2	24.7	2.2
1966	16.0	5.7	27.9	2.8
1967	18.2	7.0	31.6	3.8
1968	20.5	9.0	33.8	5.2
1969	23.1	12.1	39.7	6.9
1970	26.2	13.6	37.8	8.8

all figures in millions

Figure 18-6.   Selected Production Statistics for Some Goods Which Are a Reliable Index of Economic Growth

The Japanese people have taken to the products of industry with great vigour. Almost everyone has television, and most stations broadcast 20 hours per day. Half the time these transmissions are in colour. The quality is excellent. Vast sums are spent on entertainment, and the "neon strips" of North America are to be seen in most Japanese cities. Tourism is very popular, both within and outside Japan.

Well over 50% of Japanese households have a washing machine, sewing machine, a radio, vacuum cleaner, refrigerator, and the like. Increasingly, the Japanese dress in synthetic fabrics. There is a great demand for new and more modern gadgets of all kinds. Incredibly, there is no such thing as buying on time. A Japanese will save to buy what he or she wants; the Japanese rate of saving is the highest in the world. Since many workers' wages are often augmented by lump-sum bonuses, this helps a good deal, for such sums are banked as often as received. Higher forms of education and one's old age must still be saved for.

The Japanese are passionately addicted to the products of industry shown in Figure 18-6. Over 99% of all homes have a television set. Radio production, particularly the transistorised variety, shows a slight downturn, due per-

Japan Information Centre
Consulate General of Japan, Toronto

The Tokyo TV Tower soars 333 m into the sky. An observation platform, capable of accommodating 500 people at a time, is located at a height of 120 m. The elevator ride to the platform takes only one minute.

Japanese tourists are to be found all over the world. They throng Paris and buy art, gaze at Manhattan's famous skyline, and crowd into the South Pacific.

Japan is so westernized that it has many English words in its vocabulary. Try these!
nitah = a night baseball game
koolah = an air-conditioner
posto = mailbox
kurejito card = credit card
terebi set = television set

haps to the market being flooded. It was in the mid-1950s that the patent to produce transistors was bought from the U.S., and in 1955 Sony produced the world's first transistor radio.

The important thing to note about the products in Figure 18-6 is that they are part of the cement that helps bind the Japanese nation together. Japan's people have migrated in large numbers to the cities and are brought even closer together with such modern communications.

By 1999, it is calculated, there will be nearly 80 000 000 Japanese gathered together in the Tokaido region. This region will contain a megalopolis, embracing Tokyo, Nagoya, and Osaka. Even now, more than 60 000 000 people live in this area. Unless the Japanese do something to clean up their environment, imagine what it will be like living in Tokyo in 1999. Already there is so much industry polluting the air that suits are grimy in just a few hours and policemen on traffic duty wear gauze masks!

Japan is not just racing to catch up with the rest of the world. It has already overtaken most of it. But such rapid change is not feared in Japan; on the contrary, it is welcomed as inevitable and desirable. Even so, Japan still retains many of its traditions because they are so useful. The houses with sliding paper walls are cheap and easy to replace after an earthquake. In such a dwelling privacy is possible by the maintenance of a serenity of outlook that has been acquired over centuries of living on such a tiny archipelago, and which everyone associates with the image of Japan.

Two Japans exist. The one is modern, noisy, and very bustling. The other, the traditional Japan, is likely always to be there, however much its counterpart changes. Japan, even the world, would be the poorer for its passing.

Japan Information Centre
Consulate General of Japan, Toronto

Mountaineering and hiking are popular sports, particularly in spring.

Farming methods in Japan have undergone remarkable changes in recent years. One development is gravel culture which uses only chemically treated water and gravel. This method produces more vegetables in a shorter period of time and is not influenced by seasonal changes.

Shell Photographic Service

One thing that is still part of the traditional Japan is the social class known as *buraku-min*. The term means "hamlet people." There are 3 000 000 of them, and they are forced to live in slums as victims of prejudice. They were first stigmatized in the 16th century when Shintoism and Buddhism reached Japan. These religions hold slaughtering animals for meat and leather to be defiling, so the people who did those jobs were outcast. Gradually, *buraku-min* or *eta* (unclean people), as they were first known, came to include grave diggers, executioners, beggars, and fortune tellers. In 1969 $400 million was voted by the Japanese government to deal with the problems of buraku housing, high school scholarship awards, vocational training. But the stigma is still there, even though the caste system was declared illegal in 1871.

Japan Information Centre
Consulate General of Japan, Toronto

Japanese schoolchildren learn to write with a paintbrush.

QUESTIONS AND EXERCISES

1. What advantage does a large home (domestic market) population have for a country like Japan, which has to rely on trade to survive?
2. Write an account of the ways in which an increasingly aging population will affect the economy of Japan. Try to be specific — that is, try to pinpoint the ways in which the production of goods and services will be affected, also which goods and services will be affected. Bear in mind that if a population becomes increasingly older, the proportion of children still in school will fall. And that will affect the economy, too.
3. Draw a map of Japan and, using your atlas, mark on the major cities, railroads, and shipping routes. What do you notice about the distribution of the cities and transportation routes?
4. In what ways would the highlands of Japan be used to generate electricity? Remember that rainfall increases with altitude. Why do you think more than 70% of Japan's electrical production is produced not from hydro power but from imported oil?
5. The Inland Sea is obviously a highly industrialized region. How can you tell from the map in Figure 18-5 that this is so? Draw a map to show the extent of the Tokaido megalopolis by 1990. What will life be like in Tokaido then? Incidentally, why did Tokaido receive that name?
6. Use your local or school library to find out as much as you can about the *zaibatsu*, and in particular, how they connect with the *samurai*.

19. FISHING

SUPPORT OF LIFE

In many ways, the ocean is not treated with the attention that it deserves. It occupies 361 000 000 km² of the surface area of the world, which itself is some 510 000 000 km² in extent. The oceans, therefore, occupy 70.78% of the earth's surface. More important, however, the oceans do not merely have area. They also have depth. The average depth of the oceans of the world is 3.8 km. Thus they contain the staggering total of 1 371 800 000 km³ of water. Life of some sort or other exists throughout all that vast extent, even at the bottom of the deepest parts of the ocean where sunlight never penetrates, and where the pressure of the water is a staggering 1140 kg/cm² at its deepest known point.

People have penetrated the sea only slightly. It is not for lack of desire, of course, but simply that we labour under certain handicaps. We must use breathing aids or be encased in pressure-resistant apparatus in order to descend into the depths. Once there, we can stay only a limited time. Thus the amount of water space that has

The deepest known point in the oceans is the Marianas Trench in the Pacific Ocean. It is 11 035 m straight down.

Japan Information Centre
Consulate General of Japan, Toronto

Stern draggers such as this vessel, equipped with radar, radio and sonar, make obsolete the old-style trawler which hoisted its net over the side.

Average Height of Land 850 m

Sea Level

Continental Shelf
Maximum Depth 200 m

Lower Boundary of the
Lighted Zone of the Sea 200 m

Continental Slope

Ocean Floor
Average Depth 4000 m

Trenches

Greatest Known Depth
11 000 m

Figure 19-1.   The Principal Divisions of the Ocean

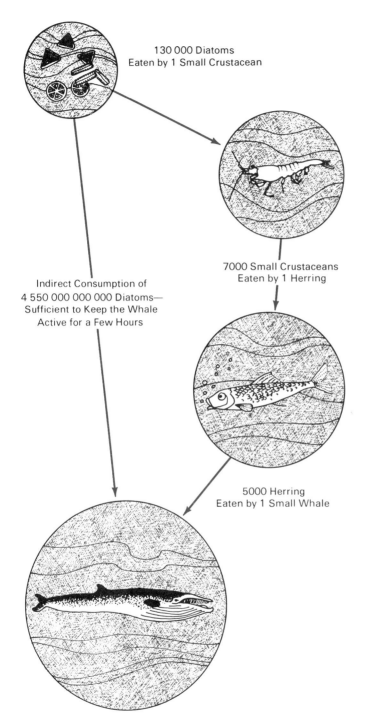

130 000 Diatoms
Eaten by 1 Small Crustacean

7000 Small Crustaceans
Eaten by 1 Herring

Indirect Consumption of
4 550 000 000 000 Diatoms—
Sufficient to Keep the Whale
Active for a Few Hours

5000 Herring
Eaten by 1 Small Whale

Figure 19-2. A Typical Food Chain

"Follow me and I shall make you [a] fisher of men." Hand nets unchanged since biblical times are still used on the Sea of Galilee.

been explored is very small indeed in relation to the whole volume of the oceans of the earth.

So far as food is concerned, mankind's major use of the oceans is for fishing. In this we are quite fortunate because the depth of the sea at the edge of the shallow, submerged portion of the continents is only 200 m and therefore accessible with nets lowered from the surface. In addition, the lower boundary of the lighted zone is only 200 m. This is the depth to which useful light penetrates.

Plankton are minute animals and plants that live in the lighted zone of the sea. The plants are very simple ones, but like all plants they need sunlight in order to assemble carbon dioxide and water to form starches and sugars. These are utilized as a food supply by the animal plankton, which are known as *zooplankton*. They eat the *phytoplankton*, the tiny plants. In turn, zooplankton are eaten by larger creatures, who are themselves preyed upon by larger creatures, and so on. This kind of connected series of animals, each one dependent upon animals lower in the series to survive, is known as a *food chain*.

The pictures of Figure 19-2 really do not tell the whole story. It is estimated that at each level in the chain, over 90% of each level is lost when it is consumed by animals of the higher level. The acts of swimming and just staying alive use a great deal of energy produced by food, just as human activities do. (Imagine what you would look like if you put on as body weight all the food you had ever eaten!) Thus the *real* consumption of diatoms by the whale is only 455 000 000. In other words, the body bulk of the whale is only $\frac{1}{10\,000}$ of the original bulk of the diatoms. The rest was used up as energy by the whale's activities. Now, if we could find some way of using *all* the diatoms . . . but we cannot, yet. We are able to utilize only the sea animals at or near the top of food chains.

Mankind's utilization of the food resources of the sea is at a very primitive stage. Most people gave up hunting animals for food a long time ago. It is easier, cheaper, more efficient, and altogether more satisfactory to domesticate animals and slaughter them as needed. So why do we not farm the sea?

In order to be able to farm the sea, we need to know a great deal more than we do now about the composition of seawater, the best concentrations of minerals for particular species of sea creatures, the best temperatures at which to keep them, the best way to make them repro-

The continental shelves of the earth occupy some 5.6% of the total area of the surface of the earth.

The baleen whales are among the largest creatures on the earth. Yet they also eat some of the smallest. They consume plankton, which they strain out of the seawater by taking a great gulp and then grinning and blowing the water between great hairy strips of whalebone that line their mouths. It would have been very hard for a baleen whale to have swallowed Jonah!

Royal Danish Embassy

A mackerel and shark auction at Esbjerg, Denmark.

The tuna is one of the few fish that swims in deep oceans across the world. Since the change-over from rods and baited lines to nets occurred in the 1960s, the world supply of tuna is in danger.

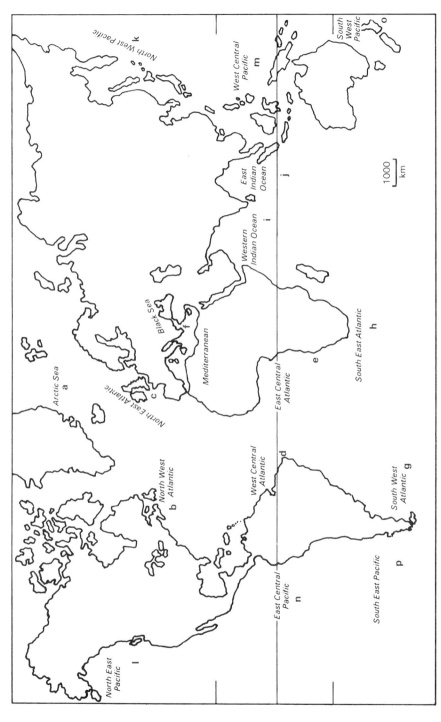

Figure 19-3. The World's Major Fishing Grounds

duce, and so on. We now use as food *thousands* of different species of fish, shellfish, crustaceans (crabs and lobsters), and marine mammals. The task of finding out everything that is necessary about them all is a stupendous one. Furthermore, someone must be willing to spend a great deal of money on developing sea-farming techniques. At the moment, the common method of catching fish by hunting them is likely to give the best return on time, effort, and money expended. But this is not to say that it will always be the case.

So far, we know the most important fishing grounds in the world are in shallow waters called *continental shelves,* because that is where most fish find their food. The bigger the area of the continental shelf, the more likely we are to find bigger concentrations of fish.

The locations of the important fishing grounds on continental shelves are shown in Figure 19-3. You may think it odd that we would show a map of the whole world in a book on Eurasia, but there is a good reason for doing so. All the fishing grounds in the world can be reached by countries in Europe and Asia. Not all countries participate, true, but some of them do. Look at Figure 19-4 very carefully.

Only the most important fish catches are listed in Figure 19-4, but several interesting things emerge. European countries do not catch a lot more fish than Asian countries. The U.S.S.R. and Japan are the only two countries to really range the world in pursuit of fish. Japan is a small country with a large population. Most of the land is mountainous, and farmland is very scarce, so if the Japanese want protein in quantity, they must go to the sea for it. The U.S.S.R. is a little different. For many years that country has had problems in trying to boost the production of meat. In the meantime, even though they are not short of land, the Soviets must also range across the oceans of the world for protein.

Generally speaking, most countries tend to fish in the oceans that are nearest to them. This is especially true of Asian countries. Except for Japan, these countries are generally too poor to be able to pay the price that fish from around the world would cost if it was caught by expensive motorized fishing vessels and packaged or frozen. Besides, the poorer Asian countries do not have that kind of technology. This is unfortunate, because these are just the countries whose inhabitants require

The U.S.S.R. has planned a new, improved factory ship of 43 000 t, twice the size of an ocean liner. It will carry 14 fully equipped trawlers, each 40 m long. The factory ship will be able to travel up to 40 000 km, and stay at sea for up to 6 months. It will be able to pack 180 000 kg of frozen fillets and 150 000 cans of seafood *daily.*

To encourage Russians to go to sea in large fishing vessels, they receive double the wage of a factory hand.

Japan Information Centre
Consulate General of Japan, Toronto

These Japanese children are well aware of their dependence on the sea for protein. Japanese people eat 10 times as much fish as they do meat. Tuna — the fish seen here — are highly regarded.

	1970 a	1972 b	1972 c	1972 d	1972 e	1972 f	1972 g	1972 h	1972 i	1972 j	1972 k	1972 l	1972 m	1972 n	1972 o	1972 p
U.S.S.R.	0.8	1 150.0	1 272.1	73.8	848.8	283.7	4.6	719.8	129.0		1 434.2	869.2		12.9	53.7	35.1
Greece						60.0										
Italy			211.7			331.7										
Poland		267.1	826.0													
Spain		233.6	255.5			91.5										
Portugal		171.0	138.7					208.7								
East Germany		142.3														
Norway			2 944.7		175.0											
Denmark		41.6	1 427.3													
U.K.			1 065.4													
France			608.5			59.9										
Iceland			726.2													
West Germany			314.4													
Japan					111.5		3.3	106.1	33.5	31.5	8 087.0	1 390.8	117.0	68.0	74.1	5.1
India									739.8	231.7						
Oman									100.0							
Pakistan									194.5							
Yemen P.D.R.									122.6							
Burma										324.0						
Indonesia										84.7			753.8			
China											2 887.0					
N. Korea											800.0				23.3	
S. Korea											1 240.5					
Malaysia													354.3		52.9	
Thailand													1 547.5			
Portuguese Timor													1 051.2			
N. Vietnam													215.0			
S. Vietnam													595.9			
Bangladesh										35.3						
Totals	0.8	4 330.0	10 710.0	1 510.0	292.0	1 090.0	760.0	2 930.0	1 730.0	780.0	14 580.0	2 740.0	4 740.0	910.0	290.0	6 220.0

Figure 19-4.   Catch by Area for Major European and Asian Fishing Countries. All figures in thousands of tonnes. See Figure 19-5 for key to areas.

	a	b	c		a	b	c
Albania	4.0	0.2	21.0	Afghanistan	1.5	0.0	7.8
Austria	2.6	1.9	50.0	Burma	446.3	4.8	4.4
Belgium and				China	7 574.0	1.3	6.3
Luxembourg	59.0	4.4	49.1	Cyprus	1.3	1.6	23.3
Bulgaria	107.9	1.2	23.8	Hong Kong	121.7	9.6	22.7
Czechoslovakia	15.1	2.0	26.7	India	1 637.3	0.6	5.0
Denmark	1 442.9	10.8	50.5	Indonesia	1 267.8	3.4	1.8
Finland	67.0	7.1	53.7	Iran	19.5	0.1	11.6
France	783.0	5.6	58.7	Iraq	26.0	0.6	13.0
E. Germany	338.2	5.4	45.3	Israel	29.3	2.7	40.4
W. Germany	418.8	3.2	50.7	Japan	10 247.8	15.5	14.2
Greece	93.3	5.8	37.2	Jordan	0.1	0.5	10.8
Hungary	27.6	0.9	41.3	N. Korea	800.0	3.6	3.5
Iceland	726.5	22.3	50.9	S. Korea	1 338.6	4.1	4.2
Ireland	92.0	2.4	56.5	Khmer	87.2	16.0	3.1
Italy	413.7	3.4	34.8	Laos	20.0	0.2	5.0
Malta	1.2	2.0	27.5	Lebanon	1.8	1.0	19.4
Netherlands	348.3	2.5	51.2	Malaysia	358.7	6.6	8.1
Norway	3 162.9	9.8	41.6	Mongolia	0.0	0.0	50.7
Poland	544.0	3.0	39.6	Nepal	2.2	0.3	6.4
Portugal	462.7	12.2	19.9	Pakistan and			
Romania	84.8	1.0	25.2	Bangladesh	459.7	0.9	9.1
Spain	1 616.9	6.3	30.3	Philippines	1 148.7	10.0	10.0
Sweden	224.7	7.8	47.0	Saudi Arabia	30.4	1.0	8.5
Switzerland	3.5	2.4	48.7	Singapore	15.7	11.3	15.1
U.K.	1 081.5	4.1	49.9	Sri Lanka	101.9	5.3	3.2
Yugoslavia	49.3	0.4	22.2	Syria	1.5	0.3	11.4
U.S.S.R.	7 756.9	3.0	32.8	Taiwan	no data	10.0	11.9
				Thailand	1 678.9	6.2	6.1
				Turkey	120.2	1.3	13.5
				N. Vietnam	300.0	3.5	5.5
				S. Vietnam	677.7	7.2	3.9
				Yemen	7.5	0.3	9.8
				Yemen PDR	122.6	19.4	14.3

a = 1972 catch in thousands of tonnes.
b = grams of fish protein available per person per day.
c = grams of other animal protein available per person per day.

Figure 19-5.   The Catches and Effectiveness of Eurasia's Fishing Fleets

proper nutrition but do not receive it. India, for example, and Bangladesh, too, are restricted in their fishing grounds to that part of the Indian Ocean that people can reach, fish and return from in a day's sailing time. (Very few of their boats have motors.)

Every human being needs animal protein in some form or other. Proteins are indispensable to a balanced diet. While the human body can *synthesize* or make up from other substances some of the proteins it needs, some it cannot. These must come from animal sources: eggs, cheese, milk, butter, meat, and fish. The first 5 things in that list are produced on land areas by man *but* they are costly, and many people simply cannot afford them. If you remember what we said about the food chains in the sea, then you will recall that only 10% of the original foodstuff an animal eats ends up as body weight. Yet

"You'll never see a gull chasing a Russian ship. Not a scrap of anything goes over the side!"— a Canadian fisherman.

animal feed all has to be paid for, and this is reflected in the price for the final animal product, meat. If, however, it is possible to catch creatures at the top of their food chains, food chains which have not had to be paid for, then this has a definite advantage. Fish are such creatures, but the cost of finding them and catching them in large enough numbers to make up for a lack of land-animal meat is an insurmountable hurdle for the poorer nations in Asia.

Figures vary, but a lot of nutritional experts believe that every person should have about 22 g of animal protein per day (on average). Allowing for the different requirements of people at different ages, this is a good working figure to apply against the average member of a country's population. You will arrive at a total animal protein figure if you add columns *b* and *c* in Figure 19-5. (Not all the countries in Asia are listed. That is because some of them are too poor to be able to afford the kind of civil service necessary to collect this type of information, even on an intermittent basis.) The map in Figure 10-1 shows the eating standards of the Eurasian countries which have collected that kind of information. If you examine the countries with a hunger problem, you will see that most of them also have a *malnutrition* problem. Not only do they not receive enough Calories per person per day (starvation), but they also do not get enough of the necessary nutrients. Fishing does not seem to help them very much. There are just too many people to be fed adequately.

Not so long ago, people thought that the sea was an unlimited source of food and that, if the malnourished people of the world were to be enabled to survive, then the ocean would provide the key. Since then some disquieting facts have emerged. The riches of the sea are not infinite, nor can they be plundered without thought if the fish harvest is to be maintained. But just because the seas belong to everybody, they are no-one's responsibility. So overfishing takes place because skippers want full holds *now*. The fishing industry in international waters is not now regulated, but it must be if fishing is to survive.

A vastly increased Soviet catch of haddock in the northwest Atlantic during 1965 resulted in a drastic catch reduction for other countries, as well as for the U.S.S.R. in succeeding years. Just imagine if that were to take place in the Indian Ocean or the West Pacific Ocean. Countries such as India, Bangladesh, or Burma would be faced with

Japan Information Centre
Consulate General of Japan, Toronto

In view of the overfishing by foreign fishing vessels on the Grand Banks, Canada started to think about claiming jurisdiction over fishing for 320 km out to sea. Iceland, Peru, and Chile already claim 320 km in their adjacent seas.

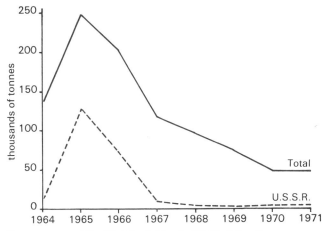

Figure 19-6.   **The Haddock Catch, N.W. Atlantic**

Japan Information Centre
Consulate General of Japan, Toronto

Alaskan fishing grounds provide a rich haul of crabs for Japan.

Many European countries eat pickled raw herring.

Raw fish in Japan is considered a great delicacy.

Fish 'n' chips is Britain's national dish. It is eaten regularly by millions of people.

disaster. Meagre as their fish catch is, it still represents a valuable addition to the people's diet.

The fishing methods of countries such as the U.S.S.R. and Japan bear a little closer examination. The kingpin of the whole operation is a factory ship, which can process, freeze, can, and pack the catch on board. The factory ship does not fish. Instead, a fleet of small trawlers or other types of fishing boat comb the seas around and take the catch to the factory ship. Then they return to catching fish again. This method is efficient and results in little wastage of fish. It is also expensive. That is why other countries are not prepared to have such fleets of fishing vessels. The cost is too great. Besides, many other European countries are able to produce most of the animal protein they need on land.

A traditional method of catching fish is by means of the *otter trawl*. For *demersal*—bottom dwelling—fish such as sole, flounder, and flatfish, the trawl is ideal. However, the trawl can destroy undersea vegetation, thus hindering recovery of the fish stock. If *pelagic* fish, those that swim near the surface, are sought, then some kind of floating net is needed. A *gill net* can be suspended from floats and held taut by weights along the bottom. A fish entangled in the mesh cannot wriggle out backwards. It tries to squeeze through, but it sticks just behind the gills, which hold firm on the mesh of the net. Large schools of surface fish can be caught in a *purse net*. A long gill net is paid out in a circle around the fish to be caught, and then a drawstring along the bottom edge is pulled tight. As the net is pulled in to the boats, the area left to the fish becomes smaller until the fish (they might be sardines) are able to be scooped or pumped out.

The Japanese have refined a technique that uses flashing lights, a long net, and a trap at the shore end. They even use cormorants to catch fish! Trained, these birds have a ring around their throat so that they cannot swallow their catch.

Possibly the most deadly method of fishing is one that utilizes an underwater light to attract fish; then an electric current paralyzes or stuns them. A large pump then sucks them up. Since all fish are taken regardless of size, breeding stocks can be wiped out by this method. Nets, on the other hand, can have different sizes of mesh so that the young, immature fish can escape to grow larger. Of course, there is little that can be done about countries that deliberately use small-size mesh nets. The International Commission on North Atlantic Fisheries (ICNAF) is paying some attention to this problem and many others like it, but the North Atlantic is only one part of one ocean and it is fished by relatively few countries.

A possible solution is proposed by Iceland. In 1972 this tiny land declared an 80 km fishing limit. Although this was ruled illegal by the International Court of Law in July 1974, the Icelanders countered with a proposal for a 320 km limit. If Iceland could enforce this, then Britain, France, Germany, and other European countries would be badly affected in the short run, but might benefit over the long run.

For the poorer countries of Asia, fishing holds much hope, but a lot of foreign aid will be needed to set up thoroughgoing fishing fleets. If all the fish caught were utilized, even the so-called "trash fish," in the form of *fish flour*, this could greatly benefit the poor nations. Fish flour is ground-up fish with the water and fat removed. When dried, the whitish substance is 80% protein. Almost odourless and tasteless, it could be added to foods as a protein supplement. As yet, it has not been put into wide-scale operation, because the people it is designed to benefit are subsistence farmers, growing their own food. They would have to add the strange substance to their own food according to directions "on the packet." But most such people are illiterate, and bound by tradition to the extent that they are suspicious of anything new. How can they be reached? They have few radios or television sets.

Inland fishing sounds a misnomer, but every Eurasian country catches fish in lakes, rivers, streams, or ponds

Different species of fish prefer different water temperatures. Hourly seawater temperature maps are produced by satellites and computers. It is hoped that this method of determining where fish are likely to be will cut down on the time spent searching for them. At the moment, up to 80% of a ship's time at sea can be spent just looking for a school or shoal worth catching.

In May 1973, the Icelandic premier, Olafur Johannesson, warned that he had instructed his gunboats to fire at British trawlers again. The following month the Iceland government said that a fleet of British trawlers had rammed an Icelandic gunboat and left it sinking. It was another incident in the "Cod War."

Caviar is fish *roe* — eggs — taken from the sturgeon of the Caspian Sea and Volga River in the U.S.S.R. It is eaten raw.

Royal Danish Embassy

Smoked herring are removed from the smokehouse, one of many on the island of Bornholm in the Baltic Sea.

especially constructed for the purpose. The rice-growing people of monsoon Asia often put small fish in the flooded paddies at the start of the growing season, and trap them as needed. It is not hard to do this, because the water is never more than about 15 cm deep.

China for thousands of years has bred carp in ponds that can be drained to allow the fish to be caught. Since the Second World War, this has been an important part of commune agriculture, because it provides protein that would otherwise be unobtainable.

Hundreds of years ago in Europe, monasteries often had fish *stews* (ponds), which provided a wide variety of freshwater fish for consumption on Fridays.

The United Kingdom spends millions of dollars annually on stocking streams and lakes with trout. To rent a stretch of a particularly good river from the person who owns it might cost a person $10 000 per week. Shellfish and crabs and lobsters are often local delicacies in Europe —for example, the oyster and clam beds at Penclawdd, Wales. In Wales, too, a variety of seaweed is gathered and eaten as a green vegetable. Called *laverbread*, it tastes a little like cabbage. Everyone must have heard of shark's-fin soup! It is made from dried sharks' fins and is a delicacy throughout the Far East. The Philippines is currently doing a roaring business selling attractive shells to foreign markets, and turtles remain a good catch for anyone in the Indian Ocean or the Central Pacific. The meat is a delicacy and the shell or *carapace* can be used to make a variety of things, from ornaments to buttons and combs.

The list of fish products is practically endless, but did you notice that at least one of the creatures mentioned above was not a fish? Turtles, of course, are reptiles. But we can count them in loosely with the fishing industry, as well as whales.

Whales are mammals but no study of the Eurasian fishing industry should fail to take account of whaling. It is a tragedy, but one with a history. Japan is believed to have the oldest whaling industry, going back to 230 B.C. By the 19th century, stocks of whales along Japan's coast had declined to the point where whaling was no longer profitable. Other countries, such as the U.S.A., had entered the industry, and by the middle of the 19th century, whaling was established in every ocean of the world. The U.S.A. was the world leader, with 735 boats out of 900 in 1846. Thereafter, with the discovery of petroleum, the U.S. share of the industry declined greatly, as petro-

Novosti Press Agency

A large sturgeon of 570 kg caught on the Caspian Sea. From such fish comes caviar.

Whales speak a language, as do porpoises and dolphins. These sea-going mammals are highly intelligent and are being researched by the U.S. Navy for a possible role in, among other things, detecting and shadowing enemy submarines. The U.S.S.R. is also conducting experiments on porpoises in the Black Sea, but "for undisclosed reasons."

leum products began to replace products made from the whale. The uses of whales included oil for lamps and lubrication, bone for ladies' corsets, whips, umbrellas, furniture, and a variety of other things.

Towards the end of the 19th century, steam whalers with explosive harpoons resulted in wholesale slaughter;

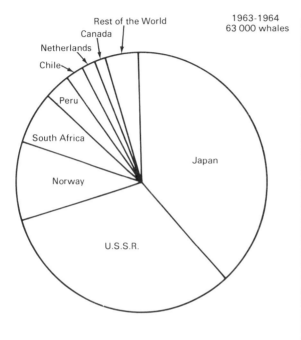

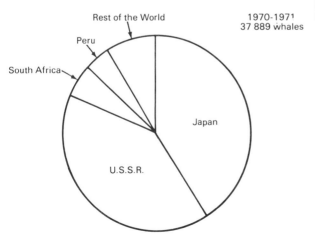

Japan Information Centre
Consulate General of Japan, Toronto

A whaling scene. The men are using long flensing knives. Very little of any part of the whale will be wasted.

Figure 19-7.   The Numbers of Whales Caught

species such as the blue whale and the bowhead were nearly wiped out. At present, the whale caught in largest numbers is the sei whale.

In the 1930s the development of factory ships allowed for the catch to expand greatly, so much so that there were two International Agreements for the Regulation of Whaling, in 1931 and 1937. In 1946, after the Second World War, the International Whaling Commission was set up. Today right and gray whales are protected; there are minimum lengths for each species, a 3-month limit to Antarctic whaling, protection for females with calves, and most of the world's oceans are closed to factory-ship whaling north of 40°S latitude. The North Pacific was left open to factory whaling as a special concession to Japan.

The IWC has voted to outlaw hunting for fin whales by 1976, and continues to restrict the number of whales that may be taken each year. In order to ensure that these regulations are followed, the two largest catchers of whales, the U.S.S.R. and Japan, have observers on each other's vessels.

In 1972 the U.S.A. proposed a total ban on the killing of all whales for at least 10 years to allow the whales to re-establish themselves. Females only calve once in 2 years, generally producing one offspring. At the moment there is a total ban on killing blue, humpback, gray, right, and bowhead whales. A vote was taken in 1972: 4 nations were in favour, 6 were opposed, and 4 abstained. Another vote on the same proposal in 1974 gave these results: 8 in favour, 5 opposed and 1 abstained. A three-quarter majority is needed, however. Maybe in 1976. . . .

The U.S.S.R. uses whales mainly for oil (lubricating oil) and grinds up the meat to feed sables on fur farms. The Japanese claim the whales provide 20% of their red meat, though this figure is disputed by the U.S., which says the real figure is perhaps 1%.

Most of the other whaling countries in the world use whale meat for pet food.

Royal Danish Embassy

Female trout are being selected to lay their eggs in another pond. This trout farm could well be copied by other countries.

	1963-64	1970-71
Blue	372	4
Fin	19 182	4 459
Humpback	318	20
Other Baleen whales, mainly Sei	13 874	10 998
Sperm	29 255	22 408
Total	63 001	37 889

Figure 19-8.   The Make-up of the Whale Catch. Actual numbers of whales are given.

What prospects are there for farming fish? Earlier in this chapter we outlined some of the difficulties, but some of them are being overcome. The Japanese actually build homes for fish to live in and are researching the potential for raising shrimp. The big advantage in farming fish is that *predation rates* on the larval stages of shellfish such as crabs and lobsters—or on any immature fish, for that matter — drop nearly to zero. In the wild, perhaps one adult emerges from 1000 fry; the rest are eaten by other creatures. Additionally, fish farms provide optimum living conditions, so that lobsters reach a marketable size in 2 years as opposed to 6 or 7 in the wild. One of the most important variables is water temperature, for it greatly affects the growth of sea creatures.

The best way to use the seas as a larder is to promote conservation and restocking programs, as well as trying to curb the rate of growth of world population. Already we use 20% of the algae and phytoplankton population of the sea in the form of fish. So we could only raise world production of fish by 4 times, and that would not be enough to feed everybody now. What will happen? Who knows? We will have to see what the future holds in store.

Norwegian Embassy Information Service

Herring from the North Sea, Norway.

QUESTIONS AND EXERCISES

1. Find out what plankton look like from your library. Draw some if you can. How large are they?
2. Find out all you can about 2 different methods of commercial fishing. Draw diagrams to show what you have discovered.
3. Use the information in Figure 19-4 to construct proportional, divided circles on a world map to show the catch in the following oceans:
    i) the entire Atlantic Ocean;
    ii) the entire Pacific Ocean;
    iii) the entire Indian Ocean.
4. On a map of Eurasia that shows the country outlines, shade in *light blue* all those countries that have an adequate supply of animal protein.
    Shade in *yellow* all those countries that have a less than adequate supply of animal protein.
    Rule *black stripes* on those countries that only have an adequate supply of protein because they fish.
    Do not forget a *legend* to show what your shading means.
5. If you look at the circles in Figure 19-7, you will see that they are proportional to the number of whales caught. Reproduce the 2 circles (outlines only), but this time divide them according to the data in Figure 19-8.

20. URBANIZATION

GLOBAL VIEW

The chances are you live in a town. Four-fifths of Canadians do. If you lived in Europe the chances are also 4 out of 5 that you would live in a town. If you lived in Asia the chances are . . . . Well, it would depend on which part of Asia you lived in; we cannot generalize about Asia in quite the same easy way that we can generalize about Canada or Europe. If you lived in Siberia the chances are high (4 out of 5) that you would live in a town. But if you lived in the Middle East or anywhere across southern Asia as far east as the South China Sea, then the chances are quite low that you would live in a town (perhaps only one out of 5, and in China maybe even as low as one out of 10). In Japan, on the other hand, the chances are again quite high (nearly 4 out of 5) that you would live in a town.

There are clearly big differences throughout Eurasia in the proportions of people who live in towns, but the differences are getting smaller as time goes by. What is happening now is that people who live in the Middle East and across southern Asia as far east as the South China Sea are gradually moving into the towns, thereby making the proportion of people who live in the towns larger. A major exception to this is China, where the people in the towns are actually being encouraged by the government to move back into the countryside. China, however, is an exception; everywhere else people are moving from the countryside into the towns. This process is called *urbanization*. It has already happened in Europe, so most people there already live in towns. There are not too many people left in the countryside in Europe to make the move into the towns; so the European proportion of 80% urban/20% rural is not likely to change very much.

In geography the word *town* is used to mean any urban area, even though some may be chartered as *cities*.

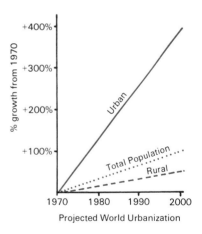

Projected World Urbanization

THE EXTENT OF URBANIZATION

Figure 20-1 gives you an indication of the distribution of the larger towns across Eurasia. You will notice that the distribution corresponds very closely with the overall distribution of population (as shown in Figure 2-1), even in the relatively unurbanized regions of South Asia and the

329

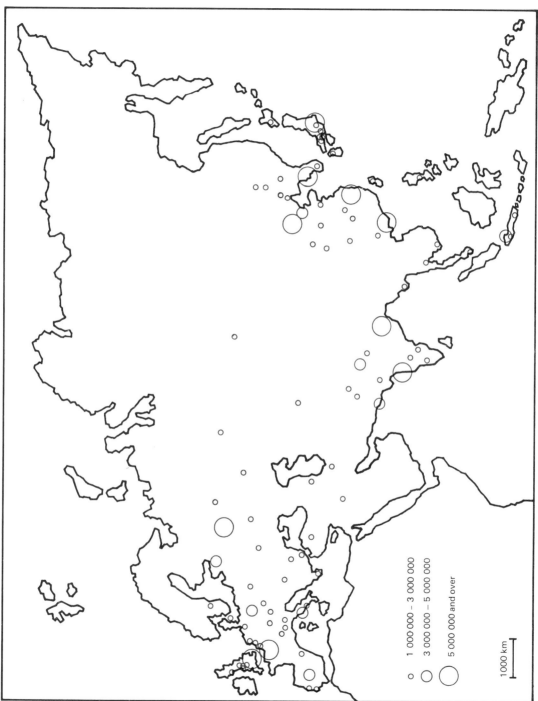

Figure 20-1. Million Cities in Eurasia (Note: Cities and Towns Are Geographically Identical)

1 000 000 – 3 000 000

3 000 000 – 5 000 000

5 000 000 and over

1000 km

Middle East — even in China. The reasons for this are obvious in Europe and Japan; most people in those areas live in towns anyway. But how are we to explain the large number of towns in India and China, where the populations are still 80% to 90% rural? The answer is easy. These 2 countries contain so many people (nearly 1.4 billion, or about 35% of the world's total population) that, even with only a low urban proportion, they are still able to contain many very large towns. It's just that the towns do not count for much in the total life of the people of those countries . . . yet.

Both Europe and Japan also had low urban proportions at one time. It used to be the normal pattern of life for most people to be farmers and to live in the countryside; only a few people lived in towns. Clearly, the need for daily food was the prime reason why so many people lived in the countryside. Farming techniques were relatively simple, productivity per person was low, and accordingly it was just about all that a farmer could do to support himself and his family with perhaps a little bit left over to support a few people in the towns. If there was nothing left over under normal conditions, then almost everyone had to be a farmer. Towns only developed when the farmers could produce at least a little bit more food than they needed to support their own families. Surplus food production was, therefore, the key to the origins of towns, both in Europe and in Japan. Even so, why should towns develop at all?

Nobody really knows the answer to the question about the earliest origins of towns. Most people who have researched the subject think that the earliest towns were probably centres for some sort of religious function — possibly worship, but also possibly tribal administration, and eventually trade. It is highly likely that early towns, in Europe and Japan, contained aspects of all these functions: a temple, a meeting-place for discussions, a place where decisions were made, a market, storage facilities for surplus food, and probably tax or tithe collectors' quarters, also with food storage facilities (a tithe barn). From this type of functional set-up it was only a short step to the addition of a military function. People who had developed temples and administrative offices and who had encouraged trade and exacted taxes were not the sort of people to allow others to take everything away without putting up a fight, so a military function was quickly part of the town's reason for existing. It was often rationalized

Rudolf Frischka

A market square in Vienna.

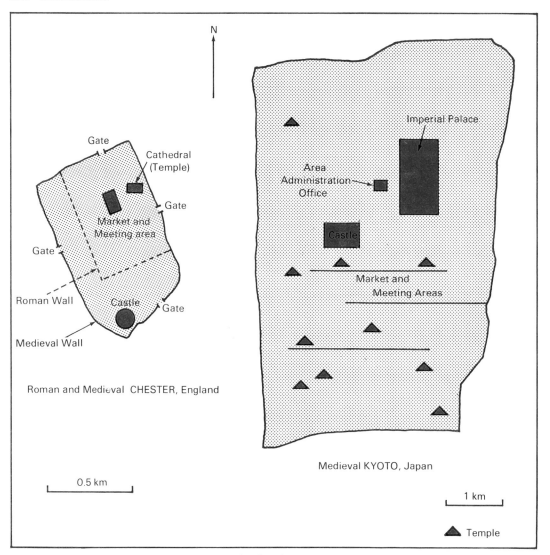

Figure 20-2. Ancient Town Functions, England and Japan

as a protective device, to protect the farmers from attack, although in both Europe and Japan, if protection were really needed from an outside attack, then all the farmers were forced to fight anyway, just as much as any of the permanent soldiers were. A much more probable reason was the need for administrators and temple people to keep the food-producing farmers in line, and to protect the town itself (rather than the farmers outside it) from various attacks. To this end a lot of the early towns had strong protective walls around them. Does it sound just a little bit as though the earliest towns in Europe and

Japan National Tourist Organization

Marilyn Olsen

(Left) A Japanese castle at Osaka, an old trading town.

(Right) A European castle near Nurenberg, Germany.

Japan were started by people whom you might call bullies? Or con men?

The situation was not very different in the early towns of South Asia. Towns here developed very early (much earlier than in Europe and Japan), and we are not sure exactly where they all were. We know that the very earliest towns started in the food-growing silt-fertilized floodplain lands of the major river systems of South Asia, be-

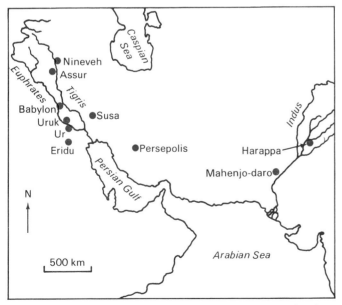

Figure 20-3. Early Towns in the Tigris-Euphrates and Indus Valley Regions

Government of India
Tourist Office, Toronto

The ghats (steps) and temples at Varanasi, where pilgrims still bathe in the Ganges.

tween latitudes (approximately) 25°N and 35°N. The chief river systems were, and are, the Tigris and Euphrates in the Middle East, the Indus and Ganges in the Indian sub-continent, and the Hwang and Yangtze in China. In the Tigris-Euphrates Valley, ancient towns such as Babylon and Ur are worth noting; in the Indus Valley there were Mohenjo-Daro and Harappa; in the Ganges Valley Allahabad and Varanasi, which still exist as religious centres; and in China, Anyang in the middle Hwang Ho Valley.

For hundreds of years the towns of Europe, Japan, the Middle East, and South Asia remained small. Rome itself, at the height of the Roman empire, did not exceed a population of 250 000 (about the same population as, say, Halifax, N.S. or Windsor, Ont.). Most towns were very much smaller than that, usually with fewer than 10 000 people. It was only the major imperial centres that could rise much above a population of about 10 000 or so, because it was only the major imperial centres that could draw upon food supplies from more than just the immediate locality. Rome, for example, drew its food supplies from as far away as Britain and Egypt. But the normal towns were restricted in size according to the amount of surplus food that could be obtained (by gift, by purchase, by trade, by exaction) from the surrounding countryside. Most towns were therefore quite small.

Throughout the time of small towns, up to the years 1800-1900, life in all countries was predominantly rural and agricultural. Towns provided a few essential services (market, worship, administration, education) and also housed a few essential trades (specialized clothing makers, tool makers, repairers of all sorts), but they were in no sense an integral part of the life of most people. Then came the revolutions.

The revolutions, agricultural as well as industrial, happened in Europe first—in England, to be exact. The agricultural revolution increased the output of food per farm-worker, so much so that over a period of time (about a century) the number of farm-workers required to work the same amount of land, *and produce 10 times as much food from it*, actually dropped in the ratio of about 5 to 1. Since the total population during the same time increased about fourfold, thus increasing demand for food by at least the same proportion, you can imagine that the agricultural revolution was a major event. You can also imagine that a lot of people ceased to be farmers; there was

A revolution is something that brings about a great change in the lives of the people. The revolution may be political, when the system of government is changed, or it may be industrial or agricultural, when the ways of earning a living are changed, or it may be something like urbanization, when the places of residence are changed.

Find out about "Turnip Townsend" and the "Coke of Holkham".

no need for them to farm. Indeed, if you put all the facts together, you can see that the agricultural revolution released a lot of labour from the farms. Let's take some figures as an example. For every 5 000 000 people in farming at the start of the agricultural revolution, there were only 1 000 000 in farming a century later; yet total population had quadrupled, so that the original 5 000 000 were now 20 000 000. The 1 000 000 in farming therefore had to try to feed this greatly increased total quantity, 20 000 000. They couldn't quite do it, for their productivity had increased only about tenfold. The difference had to be made up by Europeans buying additional food from other countries around the world. Nevertheless, the agricultural revolution in Europe was a major event.

Where did all the extra people in Europe go, now that they were no longer working on farms? Most of them, as you say, went to the towns. Quite a few, however, emigrated to the new colonies that European nations were then rapidly acquiring. Nevertheless, it was to the towns inside Europe that most of them went, looking for work. The industrial revolution introduced the idea of the factory, which was fortunate, because a factory could employ many people. However, factories that employed many people also necessitated lots of houses; so the towns that had factories in them started to grow very rapidly in population. The towns without factories (because they first lacked power supplies or raw materials) did not grow so rapidly; growth was essentially a quality of the factory town only . . . for a while. Eventually, the trading towns also began to grow, especially the ports, and, in time, the shopping towns, too. Ports were obviously fairly quick to grow after the industrial revolution had started, because they controlled much of the trade not only in imported raw materials, but also in exported finished goods. (Why did finished goods have to be exported?) Gradually, as people slowly became better off financially, shopping functions added to the growth of the towns.

During this period of the agricultural and industrial revolutions, the towns of Europe changed drastically. People flocked into them from the countryside, and the population exploded internally as well, so that from about 1800 to about 1900 a new type of town came into existence: the *industrial* town, as opposed to the *pre-industrial* town. The industrial towns of Europe had certain characteristics that made them very different from the few pre-industrial towns that still existed. They had factories; they

Many of the houses that were built to house the new factory workers were rushed up. Many of them lacked space, light, air, and piped water. Conditions were often squalid, and disease was naturally an allied problem.

The four largest towns in the world are all ports:

Greater New York	14 million
Greater Tokyo	14 million
Greater London	10 million
Greater Shanghai	10 million

had lots of workers' houses; they had a more comprehensive transportation system (initially railways, then streetcars, then buses and trucks); they tended to be dirty; and they were *much bigger*. Generally the richer people tried to escape from these rather grimy industrial cities, and so suburbs began to be developed. In time the second level of "better-off" people started to move to the suburbs, too, at which time the rich people moved even further out. Then the third level started to move to the suburbs, and so on. Eventually the town centres became the haunt of the poorest people, except during the day, when suburban people commuted to work in the offices and stores. It became quite normal indeed for the actual centre of the town to be deserted at night, no-one living there at all, because only offices and stores could afford the high rents.

The remaining pre-industrial towns of Europe (for there were *some*) didn't suffer the same sort of doughnut effect. People — usually the rich — continued to inhabit the centre of the town; there was little or no commuter traffic, the towns were clean and quiet, and they were *small*. But job opportunities were minimal. In fact, many people even left these towns in search of better oppor-

Marilyn Olsen

In central Prague the deserted streets at night indicate houses turned into offices.

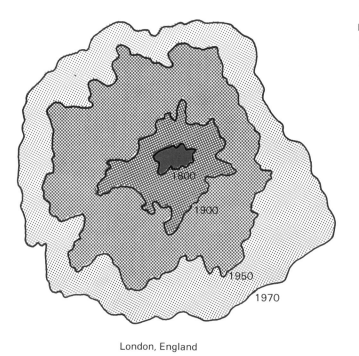

London, England

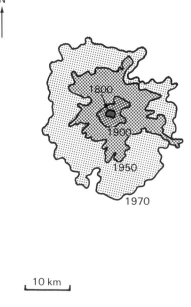

N

10 km

Moscow, U.S.S.R.

Figure 20-4.   The Growth of London and Moscow since 1800

tunities in the industrial towns. As a measure of the concentration of growth in these industrial towns, just note that at the start of the 1800-1900 period there were only 20 towns in Europe with a population of 100 000 or over; at the end of the period there were 125. This is a sixfold growth, compared with the overall population's fourfold growth. In other words, large towns were growing faster than the population in general. The population was becoming more and more urban, and less and less rural. That, of course, is what *urbanization* means.

Japan did not experience this industrial phase of urbanization until somewhat later than Europe. Industrialization in Japan was part of a deliberate policy of "Westernization" started in 1868. Agricultural and industrial revolutions of the European type were imported by the Japanese, and imposed upon the people. The results were exactly the same as in Europe; people left the land in large numbers and went instead to the towns for work in factories. The chief towns to grow as a result of this massive influx were those in the Tokaido Corridor (see Figure 20-5). Tokyo now claims that it is the largest town in the world, with a population of well over 11 million. Many people, Japanese included, also regard Tokyo as the most expensive, most infuriating to live in, most dirty, most crowded, most noisy, most exciting town in the world, too.

Industrial towns also exist in Siberia, but there they have a rather different background than the industrial

It is a common sight in Tokyo to see people standing and walking in the open air with masks over their faces. The masks are usually white in the morning — a clean one every day! — but dirty after the wearer has been out for a while. The people who wear the masks do not wear them indoors, for most of the pollution is in the streets outside.

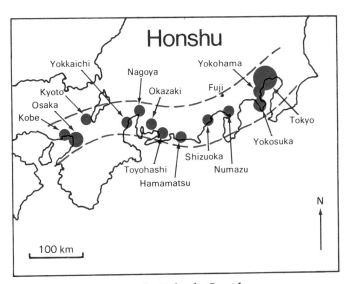

Figure 20-5. Towns in the Tokaido Corridor

Japan National Tourist Organization

Is this the most infuriating and exciting city in the world?

towns of Europe and Japan. In Siberia industrial towns were introduced largely as ready-made units, brought in from European Russia. They were intended to serve several different purposes, chiefly safety from possible attack from western Europe (especially Germany), but also colonization of an undeveloped land, development of the vast, previously untapped wealth of Siberia, and occupation of an area that might attract other interests (e.g., Chinese) if the Russians did not get there first. The Siberian industrial towns are strung along the southern edges of Siberia, where the Trans-Siberian Railway offers transportation, where the rivers out of the southern mountains offer hydro-power potential, and where the climate offers a little more ease than it would farther north. The only exceptions to the southern run of Siberian towns are the remote Arctic ports, such as Igarka and Norilsk, both inside the Arctic Circle. They are in approximately the same latitude as Inuvik and Tuktoyuktuk in Canada, but are much larger. Northern development is quite an important thing to the Russians.

The population of Siberia, then, is highly urbanized: about 4 out of 5, with the odd 1 in 5 being nomadic (such as the Tunguz in the north and the Kazakhs in the south). But it is fair to say that urbanization here was an act of

Figure 20-6. Major Siberian Towns and the Trans-Siberian Railway

colonization from European Russia, and not a local product. In other words, Siberian towns are really European
towns that have been planted in central Asia.

   Some of the towns of South Asia (as far east as the
South China Sea) are also European transplants: Hong
Kong, for example, or Shanghai; also Calcutta and Bombay; and Tel Aviv. These towns were all founded or built
up by Europeans, acting either as imperial administrators
(e.g., *New* Delhi in India, built next to the original Indian
town of Delhi, which is now still just called Delhi), or as
traders (e.g., Singapore, which was swampland before the
English came), or as settlers (e.g., Tel Aviv, which was
nothing but sand dunes before the Jews came to found a
town there). These European transplant towns were never
really a part of the life of the people in the countryside
around (except for Tel Aviv). They were where the foreigners lived. But of course, not *only* foreigners lived in
them. Many native people lived there too, as artisans,
traders, clerics, and so on. There was, therefore, *some*
drift of people to the towns, but not much.

   Not much, that is, until the Europeans generally withdrew (either physically *or* in influence). When that happened, then the local people came to regard the towns as
their own, and there was an increased drift of rural folk

One of the obvious reasons for
the local people to move into
the towns was that the towns
had been the centres of wealth,
even though the *foreigners* had
owned most of the wealth. The
association between towns and
wealth was nevertheless a
strong one, and people flocked
in because of it. Even now it is
a fact that the towns are where
the wealth is. The wealth is not
very evenly shared out, most of
it being concentrated in only a
few hands, especially in the
poorer countries.

Government of India Tourist Office, Toronto

Parliament Building, New Delhi.

Government of India Tourist Office, Toronto

Street scene, Delhi.

People flood into Calcutta.

Government of India Tourist Office, Toronto

into them. Sometimes, as in many of the largest Indian towns (especially Calcutta, Bombay, and New Delhi), the drift turned into a flood. People came from the country full of hope for work (or despair at there being no work in the countryside), and found that there was usually no work in the towns either, for the towns were not industrial towns. There was a lot of craft work, true, and many little repair shops, and many industrial-type jobs, but there were *very few factories*, and it is the factories that employ large numbers of people. If there are not many factories, then clearly there are not many job opportunities, either.

The same pattern was repeated in the non-transplants —in other words, in the locally grown South Asian towns. They were not industrial towns, either. They had their share of craft workers, small traders, repairers, and so on, but no factories. The lack of factories was crucial, for it meant that there were no large-scale job opportunities for the people who came into the towns from the countryside. However, the process of urbanization still occurred; people still moved to the towns from the countryside, but there were no jobs for them. The process was unsynchronized. Remember: In Europe the agricultural and industrial revolutions had happened more or less at the same time, and so they had in Japan, too, because the Japanese had planned it that way. But in South Asia the events were not synchronized . . . nor are they yet.

In South Asia, the 3 things that characterized the European urban revolution (namely the coincidence of the agricultural revolution, the industrial revolution, and the rapid growth of the total population) are not all present. Only two of them are — the agricultural revolution (e.g., the Green Revolution) and the rapid growth of total population. The industrial revolution is missing. Think what this means to the towns of South Asia. The farmers in the country lose their jobs, because they are no longer needed to grow crops. Nor in any event can they afford the increasing rents which the more prosperous (and innovative) farmers are prepared to pay. So they decide to go to the large town. They also decide to take their families with them, and their families are large. They cannot find work; there aren't enough factories (there are now *some,* but not *enough*). Yet they still need to eat, so they beg . . . or perhaps even steal. Eventually they may find work. The work may pay a dollar a day, and the worker who gets it is fantastically rich by local standards. More usually the work pays a dime a day, and the worker is still happy, for the competition for jobs is stiff. There is a constant stream of people coming in from the countryside desperate for work. The process of urbanization continues non-stop, and the people just pile up, waiting for opportunities; but industrialization lags.

The problem is not an easy one to solve, simply because it is so big. Millions of people are involved, not just a few thousand. Obviously, more money needs to be spent on factories. However, money spent on factories cannot then be spent on food production, and the people are already starving. If you then pay them to grow food, you cannot spend that money on building factories, and so you will never be able to provide jobs. Perhaps you could stop the people moving from the countryside to the towns —

Shell Photographic Service

The South Asian towns had their share of craft workers and repairers.

One of the big differences between European urbanization in the 19th century and Asian urbanization now is that in the 19th century the Europeans also had the new lands of Australia and North America available as an alternative destination. That option is not closed to the Asians, by any means, but it is not so wide open as it was to the Europeans 100 years ago. Thus to most Asians today the towns are the only destination possible.

There is so much unemployment and so little welfare that people will work for almost nothing. The alternatives are begging, stealing, or starving. Even a few pennies a day are better than nothing.

Hong Kong: the people just pile up waiting for opportunities.

Sue Mason

Government of India
Tourist Office, Toronto

The industrial revolution is missing.

switch off the urbanization tap? That is exactly what the Chinese have done, of course (remember, China was the exception). But the Chinese live under a totalitarian system, where the state has the power to *tell* people what to do. In India the state does not have that power; so the people still flood to the towns, and the towns cannot handle them. How would *you* solve the problem?

To be honest, you can't can you? Neither can the Indians. Urbanization there is producing problems, not solutions. Calcutta, for example, has perhaps 250 000 people out of its total population of some 10 000 000 just living on the streets: sleeping, eating, sitting around, lying about — nothing to do, nowhere to go. Millions more live in shanty towns (called *bustees* in India), which are merely put together with cardboard and empty cans. They are not too suitable when it rains, which it does every summer; India has monsoons, remember. They are not too sanitary, either, and disease is a major problem. In Calcutta there is a group of nuns, led by Mother Theresa, who do nothing but look after the wants of these poor people. The nuns hold clinics, giving injections against smallpox and leprosy; they take in abandoned infants; they bring food; they try to find jobs; they pray — sometimes they feel that is all they *can* do.

Calcutta is not alone in this type of urban poverty. All the major towns of South Asia (except China proper) have it, even prosperous Hong Kong. In Hong Kong the shanty towns rise up on the steep hillsides, precariously holding on during the heavy monsoon rains, but occasionally slipping during the frequent typhoons. The Hong Kong government is doing its best to re-house all the people in large apartment blocks, properly built, but it is a slow task,

A totalitarian system is one where the government has the power to organize the total lifestyle of all the people in the country, if it wants to. Can you think of some examples other than China?

It is estimated that by the year 2000 Calcutta will have a population of over 40 000 000 *if the present trends continue.*

5 km

■ = Bustee

Figure 20-7. Downtown Calcutta and the Bustees

and meanwhile there is a constant stream of fresh rural people from southern China (although the Peking government tries to prevent this).

Even some of the industrial towns of Europe have shanty towns around them, where the process of industrialization has gone too slowly (as in Lisbon) or the process of urbanization — the movement of people into the

People walk over the border if they can, but the communists do not let many do this. Others try to swim around the border crossings, but the communists send out patrol boats to hoist the swimmers out of the water. Sometimes the escapees try to sail a boat around to Hong Kong, but the patrol boats chase them down, too. However, the patrol boats are not always successful!

towns — has gone too quickly (as in Paris). Either way, and wherever the shanty towns occur, the processes of urbanization and industrialization are out of synchronization. The solution is somehow to bring them into synchronization, but that is easy to say and hard to do.

WHY PEOPLE MOVE TO THE TOWNS

There is no compulsion on people to move to the towns. They do it because they want to, or at least choose to. They may indeed *not want* to, but they nevertheless choose to, because the outlook in the countryside appears to them to be grim. Better hope in the town than despair in the countryside, even though the town, too, may eventually produce despair. There is hope to start with, however.

The outlook in the countryside looks grim partly because of its grinding poverty. Day after day the farmers have to work from sunrise to sunset to grow enough food for their families to subsist just above the level of starvation. Towns offer the prospects of steady hours, shorter hours, more money, greater variety of jobs, better education, better health care, more entertainment, bright lights, better housing, escape. Faced with this fascinating (if unrealistic) view of town life, many farmers decide that their future lies in the town. So they abandon their farms to the weeds, or to the farmer next door. Sometimes they are lucky. They get work, they find accommodation, they get medical care and education for their children, they make more money, they are happy. Often, however, they are unlucky, and they finish up in a shanty town.

The outlook in the countryside also looks grim because of its lack of jobs. Suppose you are a poor small farmer. The big landowners have cornered the supply of new machinery and fertilizer, and they are the only ones who can afford to install a modern pumped sprinkler irrigation system. Their crops grow much more abundantly than yours. You cannot compete in the marketplace, so you cannot pay your debts. You borrow from the moneylender, who hounds you until you decide to quit farming and go to the town to try your luck there. Your luck cannot be worse, you say. Perhaps you don't have any debts (if so, you are *very* unusual), so you decide to hang onto your farm and avoid competition in the marketplace. You will just farm your land to keep your family alive, without

Bombay has an average of one car for every 82 people. India as a whole has an average of one car for every 1131 people. In such figures lie the hopes of urban immigrants. For comparison, Canada has an average of one car for every 3 people, and Japan an average of one car for every 12 people.

Japan National Tourist Organization
A prospect of bright lights, Tokyo.

worrying about the big landowner. However, you have forgotten that the big landowner did not become big by just holding onto the land his father left him; he became big by taking over the land of other farmers. Thus he comes to you with an offer to buy you out. The offer is for more money than you have ever had before, so you take it — and take your family to the town. Maybe you even refuse the big landowner's offer. In that case the chances are that your family will argue with you to go to the town. There will be less hard work, they say, less poverty, less chance of starvation; there will be more to do, they say, more schooling for the children, more health care, more opportunity. Eventually you give in, and you all go to the town. Who made you go? What made you go?

Whatever the reason, you go. That is urbanization. The process continues.

John Bedford

Hillside shanties in Hong Kong.

Sue Mason

House boats in Hong Kong.

QUESTIONS AND EXERCISES

1. Find out what *conurbations* are. Where are some of the Eurasian conurbations? Why do you think they are there?
2. Research the history of Russian (and Ukrainian) colonization in Siberia.
3. What were the Chinese "treaty ports"? How did they come into existence?
4. Research the actual events that formed the agricultural revolution in England. You should find out about such things as the development of animal feedstuffs, especially turnips, and the introduction of different systems of crop rotation.
5. Why is the *Green Revolution* producing problems in the towns of South Asia?
6. Write an essay on the actual attractions that your town possesses.
7. How many different patterns of urbanization can you identify in Eurasia? What are they all like?